Basic Skills in Construction

Entry Level 3 / Level 1

Colin Fearn

Nelson Thornes

Published in 2011 by:
Nelson Thornes Ltd
Delta Place
27 Bath Road
CHELTENHAM
GL53 7TH
United Kingdom

11 12 13 14 15 / 10 9 8 7 6 5 4 3 2 1

A catalogue record for this book is available from the British Library

ISBN 978 1 4085 0877 0

Cover photograph/Illustration by William Howell / iStockphoto
Illustrations by Peters & Zabransky (UK) Ltd and Tech-Set Ltd
Page make-up by Tech-Set Ltd, Gateshead

Printed and bound in Spain by GraphyCems

Acknowledgements

Alamy p19 (Sara Zinelli), p21 (Roman Borodaev), p24, p25 (Studio51), p32 (Charles Stirling), p101, p102 (David J. Green); **Ancon Building Products** p65, p65; **Catnic – A Tata Steel Enterprise** p65; **Construction Photography** p10 (Buildpix); **Darren Varney / Flickr/ manfromcovorsomewhere** p66; **Fotolia** p14 (design56); p35 (Ragnarocks), p91, p101 (Julie Legrand), p124 (Kelpfish); **Getty** p63 (Dorling Kindersley); **HSE** (contains public sector information published by the Health and Safety Executive and licensed under the Open Government licence v1.0.), p2; **instant Art** p4, p4, p5, p5, p5; **istockphoto** p1, p7, p12, p13, p13, p13, p13, p13, p20, p92, p92, p92, p96, p96, p96, p98, p101, p101, p102, p102, p110, p113, p114, p116, p121; **Lie-Nielsen Toolworks** p31; **McCann Erickson PR** p95; **Photolibrary.com** p3; **Polypipe Building Products** p66; **Rex Features** p132 (Ray Tang) p133 (Design Pics Inc); **Science Photo Library** p9 (Peter Gardiner); **TWS** p132; **www.artdirectors.co.uk** p14, p33, p38, p93, p94, p96, p99, p99, p99, p99, p100, p112, p113, p114, p117, p125, p125; **www.grahambrown.com** p115; **www. rapidonline.com** p120, p121.

Colin Fearn: This book is dedicated to Helen, Matt, Tash, Days and Floyd.

The author and publisher would like to thank Heather Gunn Photography, and Clayton Rudman and the staff of Coleg Gwent, Ebbw Vale campus, for their help in producing photographs for this book.

Contents

Introduction

The construction industry requires workers who have a variety of practical skills. These skills take time and practice to master. The information in this book is designed to give you the basic knowledge to develop your skills and techniques.

The trades you will be introduced to include:

- carpentry
- joinery
- brick and block laying
- plastering
- tiling
- basic plumbing
- painting and decorating.

This book has been designed to cover the content of the City and Guilds Entry Level 3/Level 1 qualification and the practical aspects of the Edexcel Level 1 construction qualification. The City and Guilds qualification has a practical assessment at the end of each unit, and this book is designed to prepare you to pass the assessments.

Each chapter covers a group of related trades, as the materials, tools and techniques required are very similar. For instance, Chapter 2 deals with wood occupations; this includes carpentry, joinery and cabinet making.

This book is not designed to give all of the details on every trade – it is designed to provide a simple practical introduction to the skills needed in the construction industry. Each chapter introduces you to the basic materials that you will come across, such as wood, bricks, mortar and copper tube. Then you will look at tools you will need and use, and how to maintain them. Visual step-by-step guides are also provided to show you some basic techniques with photographs.

How to use this book

Key terms: explanations of trade terms that you will need to know and understand.

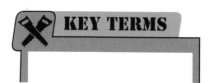

Trade tips: tips that will help you work more efficiently.

Toolbox talk: tips to help you to work in a safe manner.

Did you know: some interesting facts about the construction industry.

Have a go...: suggestions for practical pieces that you can try out to develop your skills and techniques.

Activity: activities to help you test and apply your knowledge.

Quick quiz: multiple choice questions to test your knowledge at regular intervals.

Expanding your skills: an activity that will push your ability a little bit further and gives you a bite-sized insight into what you will learn at Level 2.

Step-by-step guide: a guide to help show you how to perform techniques step by step and how to complete some projects.

STEP-BY-STEP GUIDE

Check your knowledge: at the end of each chapter, there will be some short questions that will test your knowledge and understanding of the chapter you have just completed.

Remember: important points to remember.

Health and safety

It is very important to keep safe and healthy when working. There are many dangers in the construction industry. You could have an accident that could injure you, or you might come into contact with something that can damage your health. If you pay attention to keeping safe and healthy, you have a good chance of enjoying a long career in the building industry free from injury.

The construction industry has become a much safer place to work in recent years. However, by not taking enough notice of staying safe and healthy, thousands of people are injured or even killed each year on building sites and in workshops.

Many more suffer from health problems caused by work, for example bad backs, lung damage, deafness, skin problems or other health issues such as vibration white finger. Therefore, it is important that we talk about the subject of health and safety first.

In this chapter you will learn about:

- ■ UK health and safety law
- ■ some of the dangers in the construction industry
- ■ reducing the risk of injury to ourselves and others around us
- ■ what to do if an accident occurs.

HASAWA: Health and Safety at Work Act – the law in the UK that deals with health and safety.

COSHH: the Control of Substances Hazardous to Health.

DID YOU KNOW

During 2009 and 2010, there were 41 fatal accidents in the construction industry within the UK (source: HSE).

UK health and safety law

In the UK, there are many laws that have been put into place to make sure you and those around you are safe when at work. If these laws are not obeyed, then there is a greater risk of injury and damage to yourself and others. You could also lose your job or be taken to court.

This is a brief introduction to two of the main laws that you need to obey: the Health and Safety at Work Act (**HASAWA**) and the Control of Substances Hazardous to Health (**COSHH**).

Figure 1.1
HASAWA notice

The Health and Safety at Work Act 1974 (HASAWA)

HASAWA applies to all workplaces. It covers everyone involved including the workers (whether they are employed or self-employed), the employer, subcontractors and those delivering goods to the workplace. It protects those who are working at a workplace and anyone else who might be affected by the work.

HASAWA advises employers and employees to follow certain rules in order to secure their health, safety and welfare at work. These rules are discussed below.

Employers must make sure that it is safe to enter and exit the work area. The work area must be safe to work in and this includes all machinery. Employers must carry out regular risk assessments to ensure that there are minimal dangers to their employees in a workplace. **Personal protective equipment (PPE)** will be provided free of charge by the employer to their employees. Employers must ensure that the appropriate PPE is worn when needed.

Employees and **subcontractors** must work in a safe manner. Not only must they wear the PPE that their employers provide, they must also look after the equipment. Employees must not be charged for anything given to them or done for them by the employer in relation to safety.

Figure 1.2 A construction worker wearing PPE

KEY TERMS

Personal protective equipment (PPE): equipment worn to protect the user from things like dust, sparks and splinters.

Subcontractors: workers not directly employed by the contractor, often known as 'subbies'.

The Health and Safety Executive (HSE) (the national independent watchdog for work-related health, safety and illness) ensures that the law is carried out correctly and can make spot checks in the workplace. The HSE can bring the police with them, examine anything on the premises and take things away to look at.

If the HSE finds something that is a health and safety problem, they might issue an improvement notice giving the employer a set amount of time to correct the problem. They can also issue a prohibition notice that will stop all work until the problem is put right. They might even take an employer, employee, subcontractor or anyone else involved with the building to court.

Safety signs in the workplace

The law also sets out the types of safety signs we must all take notice of. There are different signs that warn us of danger, and tell us what to do in order to keep safe. You need to know that there are five basic types of sign.

■ Prohibition signs – these signs are round and red and white in colour. They tell you **not** to do something.

Figure 1.3 A prohibition sign

■ Caution signs – these signs are usually triangular in shape and yellow and black in colour. They warn you of hazards.

Figure 1.4 A caution sign

■ Mandatory signs – these signs are round and blue in colour. They tell you that you **must** do something.

Figure 1.5 A mandatory sign

■ Safe condition or information signs – these signs are usually square and green in colour. They can tell you the safe way to work or important information.

Figure 1.6 A safety sign

■ Supplementary signs – these white signs give you extra important information.

Figure 1.7 A supplementary sign

Control of Substances Hazardous to Health (COSHH)

The COSHH law controls the use of dangerous substances, such as preservatives, fuel or oil-based paints. These substances need to be moved, stored and used safely without polluting the air. COSHH also covers hazardous things produced during work, such as wood dust through sanding or drilling. There are other things covered by this law, such as dust from sand or bacteria.

QUICK QUIZ

What does COSHH stand for?

a) Continual Oversight Saves Hazardous Handling

b) Controlled Only Substances and Healthy Hazards

c) Control of Substances Hazardous to Health

d) Control of Substance of Healthy Home

Dangers in the construction industry

 ACTIVITY: IDENTIFY

Look at Figure 1.8 below. How many dangers can you spot? Would you like to work in this workshop?

Figure 1.8 Spot the hazards

Fire

Fire needs three things to burn:

- ☐ Oxygen: a gas in the air that reacts with flammable substances.

- ☐ Heat: a source of fire, such as a match or hot spark.

- ☐ Fuel: flammable things such as wood, petrol or paper.

If you have oxygen, heat and fuel, you will have a fire.

Figure 1.9 The fire triangle

How to avoid the outbreak of fire

Being tidy with offcuts and materials will help to prevent a fire from starting. Waste should go into proper bins or skips. Smoking around flammable substances should be avoided. Dust can be very flammable and can even be explosive, so it is important to use dust bags, extraction and have good ventilation in workshops. Flammable liquids, such as oil-based paints or thinners, must be locked away in a metal cupboard.

What to do if there is a fire

If a fire does break out, it is important to raise the alarm and calmly exit the building, making your way to the assembly point. Don't stop to get your things. When you reach the assembly point, your supervisor will make sure that everyone has got out of the building.

Figure 1.10 An untidy work area can be hazardous

Types of fire extinguisher

For tackling small fires, it is crucial to use the right sort of extinguisher for the fire you are fighting. For example, putting water on an oil fire could make it explode. See the table below for the different types of fire extinguisher.

Type of fire risk	White label	Cream label	Black label	Blue label	Yellow label	Red
	Water	Foam	Carbon dioxide	Dry powder	Wet chemicals	Fire blanket
A	✓	✓	✗	✓	✓	Can be used for smothering all types of fire. Also for use where clothing is alight since it does not pose a risk to skin or to breathing as some extinguishers do.
B	✗	✓	✓	✓	✗	
C	✗	✗	✓	✓	✗	
D METAL	✗	✗	✗	✓	✗	
E	✗	✗	✓	✓	✗	
F	✗	✗	✗	✗	✓	

Figure 1.11 Fire extinguisher table

Lifting

Another risk to your health while you work is not lifting things in the correct way. You can easily damage your back if you bend in the wrong way or if you overstretch.

Here are a few things to think about when lifting:

◼ Take a look at the load first. Is it too heavy? Is it a difficult shape? Do you have to stretch to reach it? If so, get some help.

- Where are you going with the load? Is the path clear and is there somewhere to put it down?

- Lift with your back straight, elbows in, knees bent and feet slightly apart.

- Put the item down carefully; watch out that your fingers do not get trapped!

- Make sure that the stack of materials you are making isn't unstable or it might topple over.

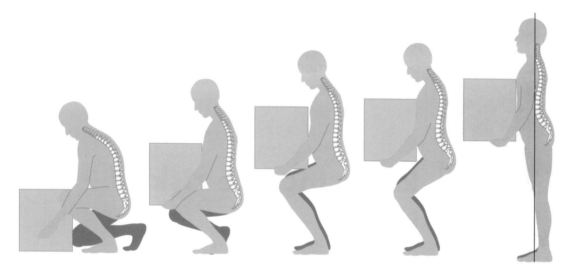

Figure 1.12 Lifting correctly

Faulty tools

Tools can be in a dangerous condition. They might be blunt, or have 'mushroom' heads (see Figure 1.13 showing a mushroomed head on a cold chisel). Hammers sometimes have loose heads or broken handles. Tools that have faults like these need to be fixed before they can be used again. Keeping tools in top shape is discussed further in each chapter.

Figure 1.13 Do not use dangerous tools

Working platforms

Sometimes you need to stand on something to reach the work to be done. This is known as a working platform or access equipment.

It is important to use safe, well-maintained equipment that is designed for the job. Ladders should be free from damage, such as missing rungs or splits in the sides. Steps should always be used on a level surface.

You will need to be trained in putting up and using access equipment. This is discussed in more detail in Chapter 5 on page 121.

Figure 1.14 A working platform

Reducing the risk of injury

It is important to reduce the risks we find when working. PPE is very important, but it is just the last line of defence; it protects you if there is an accident. It is much better to try and prevent the accident in the first place. A lot can be done to make the workplace a safer place.

KEY TERMS

Risk assessment: a form that is filled out highlighting all the risks involved with a certain job and how to deal with those risks.

The employer will make sure a **risk assessment** is completed covering every task to be carried out. This looks at each risk, and how that risk can be reduced or even removed completely. Working in a safe way also reduces risks. You need to be properly trained. You need to work in a tidy way. It is important to make sure that your tools are kept sharp, as blunt tools need a lot of force to use and can cause accidents. Do not cut towards your body or hands. There will be a lot of safety tips like this as you go through this book.

Despite taking care to ensure that your work area, equipment and tools are safe, these could still hurt you even if you use them in a safe way.

REMEMBER

You need to be aware of dangers, and some dangers can be put right on the spot. Sometimes you will need to tell your supervisor that you have spotted something that is not safe. If you spot a danger, do something about it!

BBS Construction Services

RISK ASSESSMENT

Activity covered by assessment: _____

Location of activity: _____

Persons involved: _____

Dates of assessment: _____

Tick appropriate box ☑

- Does the activity involve a potential risk? YES ☐ NO ☐

- If YES can the activity be avoided? YES ☐ NO ☐

- If NO what is the level of risk? LOW ☐ MEDIUM ☐ HIGH ☐

- What remedial action can be taken to control or protect against the risk?

1 _____
2 _____
3 _____
4 _____
5 _____

MANAGEMENT SUMMARY:

Priority for action: LOW ☐ MEDIUM ☐ HIGH ☐

Action to be taken: _____

Date action to be taken by: _____

Date for reassessment: _____

Assessor's name and signature: _____

ASSESS THE RISK – PUT IN CONTROLS – CHECK THEY WORK

Figure 1.15 Example of a risk assessment form

Types of PPE

PPE is a very important part of staying safe. There are different types of PPE that are designed to protect different parts of your body.

 ACTIVITY

In pairs, consider the following two scenarios:

a) You are using an oil-based paint that has a strong smell.

What would be the best way of controlling this risk? What would you do?

b) You spot a board thrown away on the floor with nails sticking out of it.

What would be the best way of controlling this risk? What would you do?

TOOLBOX TALK

Always wear a hard hat in the right manner. Do not wear it back to front; it is designed to fit your head one way only and will not protect your head if fitted incorrectly.

Hard hat or safety helmet

This is to be worn when there is a danger of objects falling and hitting your head. Most building sites you work on will require you to wear a hard hat or safety helmet. Be sure to adjust the headband correctly, so it fits your head well, and do not wear any other hats underneath it.

Figure 1.16 A hard hat

Safety boots or shoes

These should be worn when there is a danger of dropping tools or materials on your feet. You would need to wear these for any work in construction. Some safety footwear has sole protection to help prevent nails going up through your foot if you stand on one. They will also have toe-caps, which can be steel or hard plastic.

Figure 1.17 Safety boots

Ear defenders

Your ears can be very easily damaged by loud noise. Ear protectors or ear plugs will help to prevent hearing loss while using loud tools or if there is a lot of noise around you.

REMEMBER
You only get one pair of ears, so look after them!

Figure 1.18 Ear defenders

High-visibility jacket

The bright reflective colours of a high-visibility ('hi-viz') jacket ensure that you are easily seen by others, so you do not get hit by site traffic.

Goggles or safety glasses

It is easy to get something in your eye when working in construction. You might get small pieces of metal, wood or dust in your eye. This is painful and could damage your eyesight. You only get one pair of eyes, so protect them with goggles or safety glasses.

Figure 1.19 Hi-viz jacket

TRADE TIP

Goggles tend to steam up, especially if you use them with a mask. You might find safety glasses easier to use.

Figure 1.20 Safety glasses

ACTIVITY

Name five different items of PPE. List the jobs each one may be suitable for.

Face mask

Dust can damage your lungs and cause difficulties with breathing. Dust is often made during construction and is a very common danger to your health. A face mask will prevent you from breathing in the dust.

Figure 1.21 Face mask

In some cases, you will need a special mask to keep out fumes from paint or preservatives. Your supervisor will provide you with the mask you need.

Gloves

Gloves protect your hands when there is a risk of splinters or getting hazardous substances on your hands.

Figure 1.22 Safety gloves

What happens if there is an accident?

If you have an accident, it is important to let your supervisor know immediately. The accident will be recorded in the accident book. If needed, you will be given first aid by a first aider. First aiders are specially trained; they have to attend regular training to keep up to date.

The law states that all accidents must be recorded, and any accident that results in more than three days off work must be reported to the HSE.

Even minor cuts could become infected, so it is best to get any injury checked out if it becomes red or inflamed.

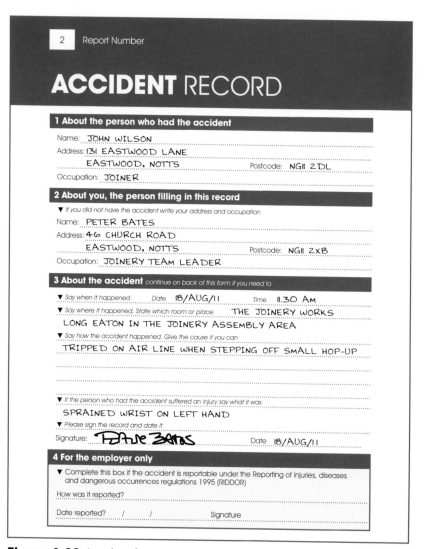

Figure 1.23 Accident form

IN THIS CHAPTER you have learned about:

- the health and safety law in the UK

- risks found in the construction industry

- how to reduce the risk of being injured

- what to do if you get injured.

This is only an introduction to health and safety and you can read more in *A Building Craft Foundation*, 3rd edition by Peter Brett. While you are carrying out the projects covered by this book, be sure to stay safe and healthy.

CHECK YOUR KNOWLEDGE

REVISION QUIZ

1. What three things are required for fire?

2. Which sign tells you not to do something?

3. Which sign warns you of hazards?

4. Draw a mushroomed head on a bolster.

5. List three things you can do to lift safely.

6. Which law deals with the storage of hazardous substances?

7. Which law deals with general health and safety?

8. Name three items of PPE and explain what each is used for.

9. What is a risk assessment?

10. List two reasons why it is important to keep a tidy workspace.

Wood occupations

2

Timber is a natural product that has been used to build with for thousands of years. It is attractive, fairly easy to work with and is very kind to the environment if it is grown in a sustainable way.

Wood occupations include three trades:

- Carpentry
- Joinery
- Cabinet making.

We will cover each of these trades in more detail in this chapter, as well as looking at some of the tools that you will need to recognise and know how to use.

In this chapter you will learn about:

- the different wood trades
- woodworking materials
- woodworking tools
- tool maintenance
- techniques.

The information in this chapter covers skills and knowledge to complete units: 001, 002, 003, 101, 102, 103, 104, 105, 106, 107, 201, 202 and 203.

Wood trades

People who work in the three main wood trades include:

- carpenters
- joiners
- cabinet makers.

Carpenters

KEY TERMS

First fix: the carpentry jobs that are carried out before plastering.

Second fix: the carpentry jobs that are carried out after plastering.

Carpenters generally build with wood. Building a house has two main stages for a carpenter and these are called **first fix** and **second fix**.

Generally, first fix jobs include:

- erecting timber frames
- installing joists and rafters
- fixing door frames and door linings.

Second fix jobs include:

- hanging doors
- fitting locks
- fixing skirting and architraves.

Joiners

Joiners build items that carpenters will fit, such as:

- ☐ doors

- ☐ windows

- ☐ stairs

- ☐ handmade kitchens

- ☐ decorative mouldings (such as skirting boards and architraves).

Figure 2.1 Architrave being fitted

Cabinet makers

Cabinet makers make furniture. These items can include:

- ☐ tables

- ☐ chairs

- ☐ chests of drawers

- ☐ cabinets

- ☐ boxes.

 **ACTIVITY**

List three first fix and three second fix carpentry jobs.

Woodworking materials

Wood

Wood is basically divided into two types: **hardwood** and **softwood**. These terms have nothing to do with whether the wood is hard or soft, but whether the wood comes from deciduous trees (ones that have broad leaves) or coniferous trees (ones that have needles and cones).

Different types of timber are used in construction. Softwood is most commonly used in the building industry; timber from spruce trees (whitewood) is used for general construction, while timber from redwood trees is used for joinery work. Softwood is a strong, lightweight material with a very distinct grain.

A tree puts on an extra ring of growth every year. Each ring consists of two colours – the darker area of the ring is slower winter growth, and the lighter area is the faster summer growth. The dark areas are harder than the light areas between them. The wood can have **knots** in it that can affect its strength and workability.

Figure 2.2 Cross-section of a softwood tree, showing the summer growth (lighter areas) and the winter growth (darker areas)

In whitewood, the lighter area (summer growth) is very soft. This can cause problems if your chisel is not very sharp, as the wood can bruise instead of cutting properly.

Wood can be made into panel products, which are available in larger widths than standard timber. These products are more stable (they do not warp or bend) and are ideal for panels, furniture and other applications where wide, flat timber sections are needed. There are many types of panel products, two of them being plywood and medium density fibreboard (MDF). Plywood is made from many thin layers glued together; it is a very stable material. MDF is made from wood fibre and has a very smooth surface.

Other types of material

Adhesives

Adhesives (glues) are used by the woodworker. The one we will consider in this book is the most commonly used – polyvinyl acetate (PVA). PVA is water-based, strong, has no smell and is non-toxic. After the assembly of a joint or frame, waste glue is easily removed using a damp rag.

Screws

Screws are commonly used for fixing timber. They are often made of steel and are usually coated with zinc so they do not rust. Screws are also available in other metals, such as brass. There is quite a variety of

Figure 2.3 Countersunk screws

screw heads available; these are discussed later in this chapter on page 37. The screws shown in Figure 2.3 on the previous page are countersunk screws. This is the most commonly used screw.

Nails and pins

Nails and pins are commonly used to fix timber. Pins are smaller than nails and are used to fix smaller sections, such as beads.

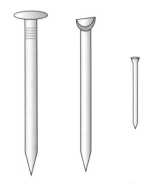

Figure 2.4 Round wire nail, oval nail and panel pin

Woodworking tools

There are many tools available to the woodworker. The tools can be grouped as follows:

- marking-out tools
- chisels
- planes
- saws
- striking tools
- drills and screwdrivers.

Marking-out tools

Marking out involves transferring the dimensions from the **rod** or drawing to the work piece. It is important that the marking out is accurate. If it is wrong, then the job will be wrong, no matter how good your cutting is.

KEY TERMS

Rod: a full-scale workshop drawing showing all the required detail.

Therefore, it is important to use the proper marking-out tools and to keep them maintained.

Figure 2.5 shows the main tools for marking and setting out:

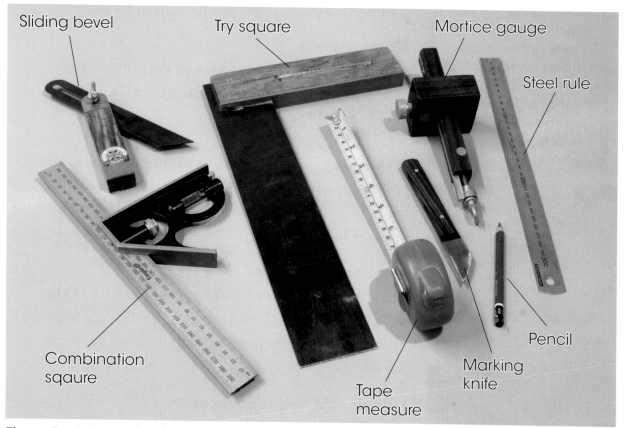

Figure 2.5 Selection of marking-out tools

Pencil

The pencil could be called the most important tool when marking out. The pencil needs to be sharp and not too soft. There is a range of pencils available.

TRADE TIP

2H pencils are best for marking out joinery on to planed timber, while HB pencils are best for marking out on to sawn timber. Using a pencil that is too soft on an item of joinery will result in a lot of smudging and the job will become dirty.

Figure 2.6 A try-square in use

Figure 2.7 A combination square in use

There are soft 'B' pencils, which range from B to 4B. These pencils are for art use and are not suitable for carpentry. Hard pencils range from H to 4H. 2H pencils are ideal for drawing and marking out joinery. 4H pencils are too hard. The most common pencil is HB. The 'lead' in this pencil is neither hard nor soft and these pencils are ideal for carpentry work.

Marking knife

A marking knife is used to mark very fine lines, especially for fine work such as dovetails.

Try-square

A try-square is a tool used to square around timber. Notice how the stock (the thick wooden bit of the square) is kept on the face without sticking out past the edge of the wood.

Combination square

A combination square can be used to mark 90 or 45 degree angles. It can also be used as a gauge by adjusting the blade and running it along the edge of the timber.

Folding rule and tape measure

A folding rule is used to take measurements. It can be made of steel, plastic or wood. Retractable tape measures can also be used.

Gauge

A gauge is used to mark parallel lines down the length of timber. There are several types of gauge: the marking gauge (which has one pin and is used for marking housings); the mortice gauge (which is used for marking out mortices); and the cutting gauge (which can be used for marking out hinges).

Figure 2.8 A mortice gauge

Sliding bevel

A sliding bevel is used to mark angles, such as dovetails. A chalk line can be stretched between two points and then flicked to produce a straight line.

 ACTIVITY: IDENTIFY

Look at the marking-out tools below. Identify each tool and explain its use.

A

B

C

D

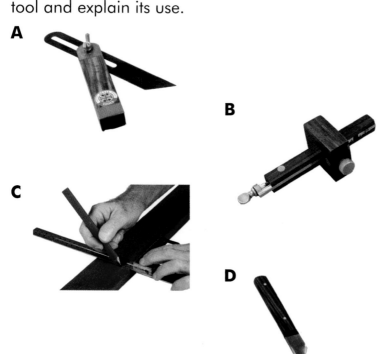

 TRADE TIP

The building industry uses either millimetres (mm) or metres (m) when taking measurements. However, most rules and tapes are marked in centimetres (cm), which can be confusing. To convert cm to mm simply multiply by 10. For example, 2.5 cm is the same as 25 mm.

 QUICK QUIZ

Which is the best pencil to use for marking out joinery?

a) 2H

b) 2B

c) 4B

d) 4H

TRADE TIP

When hitting a chisel, it is best to use a wooden mallet rather than a hammer. A mallet is heavier and will make morticing easier; whilst a hammer is likely to damage the handle of the chisel.

Chisels

A chisel is a very important tool for anyone working with wood. A chisel is used for chopping mortices, cutting out housings and for generally cleaning up joints before assembly. It is made from a long blade and the handle can be made of wood or plastic.

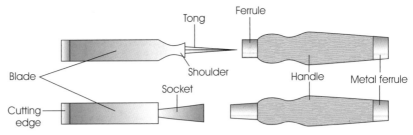

Figure 2.9 Parts of a chisel

Types of chisel

Bevel-edged chisel
A bevel-edged chisel is the normal everyday chisel in common use. The long edges are bevelled, making it easier to get into tight corners, such as a dovetail.

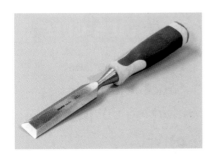

Figure 2.10 Bevel-edged chisel

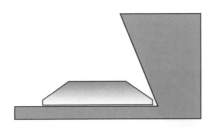

Figure 2.11 A bevel-edged chisel will cut into tight corners

Firmer chisel
A firmer chisel is square in section, not bevelled. This means it cannot be used in tight corners like the bevel-edged chisel. However, because there is more metal in the blade, the firmer chisel is stronger.

Figure 2.12 Firmer chisel

Mortice chisel

A mortice chisel is very thick and strong. This means it can withstand a lot of hitting and sideways force. It is used to make mortices

Figure 2.13 Mortice chisel

in timber and to form **mortice and tenon joints**. The sides of the blade are deep, so it follows the side of the mortice and this stops the chisel twisting in use.

Franked mortice and tenon joint

Hauncheon or franking

Haunch

Moulded frame mortice and tenon joints

Scribed moulding

Secret haunched joint

Tapered haunch

Mitred moulding

Figure 2.14 Mortice and tenon joints

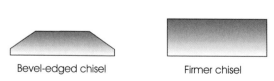

Bevel-edged chisel

Firmer chisel

Mortice chisel

Figure 2.15 Cross-sections of chisel types

QUICK QUIZ

Which of the following would be the strongest chisel?

a) Mortice

b) Firmer

c) Bevel-edged

TRADE TIP

Use the right tool for the right job. Using a tool that is not designed for the task could break the tool, and is often dangerous!

Planes

A plane is used to smooth and shape timber to leave a flat surface. There are many types of plane. Each type has a different function, but most of them have the same parts.

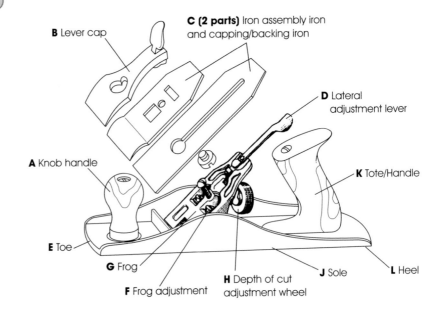

B Lever cap

C (2 parts) Iron assembly iron and capping/backing iron

D Lateral adjustment lever

A Knob handle

K Tote/Handle

E Toe

G Frog

F Frog adjustment

H Depth of cut adjustment wheel

J Sole

L Heel

TRADE TIP

When replacing the backing iron, do not leave too big a gap between the cutting edge and the backing iron (no more than 1 mm). The smaller the gap, the finer the cut.

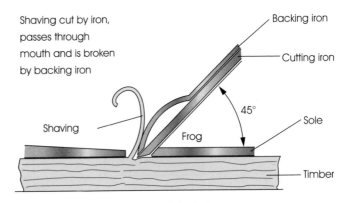

Shaving cut by iron, passes through mouth and is broken by backing iron

Backing iron

Cutting iron

45°

Sole

Shaving

Frog

Timber

Figure 2.16 A plane with the parts labelled

How to set up a plane

You will need to break down a plane to check the condition of the iron and re-sharpen if necessary.

A Removing lever cap

B Removing backing iron

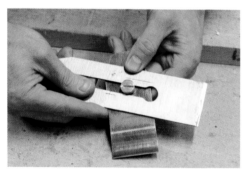

C Replacing backing iron. Note how the backing iron is placed at 90 degrees and then turned. This prevents damage to the newly sharpened edge.

D Replacing capping iron. If the backing iron is a bit loose or tight, adjust the screw as shown.

E Use of adjustment wheel

F Use of adjustment lever. Note that the blade will move in the opposite direction to the lever.

Types of plane

As there are different types of plane available, it is important to choose the correct plane for the job. Some planes are better designed to produce flat surfaces and others may be used for cutting rebates, grooves and bevels.

Smoothing plane

A smoothing plane is a short plane used for cleaning up work. It is not very good at planing straight faces or edges, as it will follow an uneven shape.

Jack plane

A jack plane is a longer plane and is good for planing straight. However, it is not very good at cleaning up work, as the long bed will ride on the high spots.

Try plane

A try plane is very long and is excellent at making timber straight. It is a very expensive plane and difficult to carry around, so you do not see these planes very often.

Figure 2.17 Smoothing plane

Figure 2.18 Jack plane

Figure 2.19 Try plane

Block plane

A block plane is a very short plane that is generally used for cutting end grain. This plane has a different angle of cut and no backing iron, so it can rip the grain if used for cleaning up work.

Figure 2.20 Block plane

Rebate plane

A rebate plane is used to produce **rebates**. Some planes have two fences (guides) that can be set to the required sizes of the rebate.

Figure 2.21 Rebate plane

Figure 2.22 Duplex rebate plane

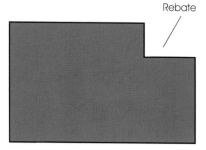
Rebate

Figure 2.23 Diagram of rebate

Spokeshave

A spokeshave is used to plane curves. There are two types: one for internal (concave) curves; and one for external (convex) curves.

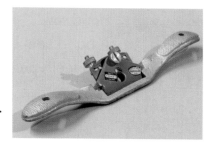

Figure 2.24 Spokeshave

Figure 2.25 Spokeshave beds

Figure 2.26 Hand router

QUICK QUIZ

Which is the best plane to use for cleaning up joints?

a) Jack plane

b) Smoothing plane

c) Try plane

d) Plough plane

Hand router

A hand router is used to produce housings. It should not be confused with an electric router.

Plough plane

A plough plane is used to cut grooves. It is fragile and quite expensive, so great care needs to be taken when using one. Combination planes are very similar. Grooving is now very often done with an electric router.

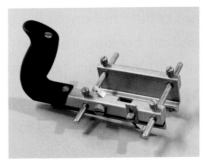

Figure 2.27 Plough plane

Saws

Saws are used to cut wood and there are many types. They are made up of a blade and a handle.

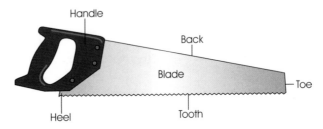

Handle

Back

Blade

Toe

Heel

Tooth

Figure 2.28 Parts of a saw

Types of saw

Hand saw

A hand saw is a general-purpose saw with 5–8 teeth per 25 mm of blade.

Figure 2.29 Hand saw

Panel saw

A panel saw is very similar to a hand saw, but with smaller teeth. It has 7–12 teeth per 25 mm of blade.

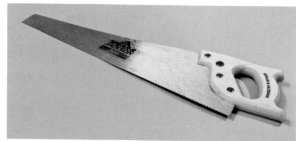

Figure 2.30 Panel saw

Tenon saw

A tenon saw has a back attached to the blade to make it stiffer. A tenon saw is used for cutting joints and is very good at cutting across the grain of the wood. It is used with a **bench hook**. It has 12–14 teeth per 25 mm of blade. A dovetail saw is a type of tenon saw with smaller teeth (18–24 teeth per 25 mm).

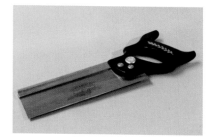

Figure 2.31 Tenon saw

Figure 2.32 Bench hook

Coping saw

A coping saw is used for cutting curves. It is important to keep the blade **tensioned** correctly and to cut carefully as the blade can snap quite easily.

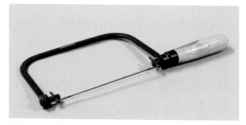

Figure 2.33 Coping saw

KEY TERMS

Bench hook: this is used to hold timber on the top of the bench whilst cutting shoulders.

Tensioned: when a blade is kept tight. With a coping saw, this is done by turning the handle.

QUICK QUIZ

What is a tenon saw best suited for?

a) Cutting curves

b) Chopping mortices

c) Cutting joints

d) Smoothing timber

☼ ACTIVITY: IDENTIFY

Name the three tools below and explain what each one is used for.

A

B

C

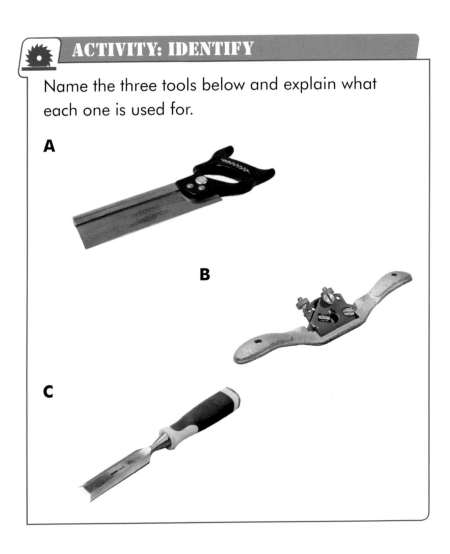

Striking tools

Hammers and mallets are commonly used by the woodworker. Hammers are used to drive fixings and mallets are used for assembly and with chisels.

Hammer

A hammer is used to drive nails, pins and wedges. A hammer is not for driving screws! Some hammers have a claw; this can be used to pull out nails.

Figure 2.34 Selection of hammers

Mallet

A mallet is made of hardwood and is used to strike chisels when housing or morticing. A mallet is also useful for the assembly of joints, as it does not damage the wood.

Figure 2.35 Mallet

Drills and screwdrivers

A drill is used to make holes in wood. There are several types of drill bit, which fit into the **chuck** of the drill. A screwdriver is used to drive screws into wood and give a very good fixing. A battery drill can drill holes and drive screws.

A wheel brace is used for drilling small holes.

Figure 2.36 Wheel brace

A swing brace is used for drilling larger holes.

Figure 2.37 Swing brace

A battery drill is ideal for pre-drilling and driving screws. If it has gears, first gear is best for screws and the second gear is best for drilling. Many drills have a clutch to make sure that screws are not over-tightened, causing them to snap, **cam out** or vanish below the surface.

Figure 2.38 Battery drill

Figure 2.39 Clutch and gear settings

Types of drill bit

Auger bit

An auger bit (centre bit) is excellent for drilling larger holes in wood. Notice the 'worm' at the tip of the bit. This pulls the bit into the wood. Be careful not to drill right through though, as you will split the wood on the other side. When you see the worm starting to come through, stop drilling and drill from the other side.

Figure 2.40 Auger bit

Figure 2.41 Split out when drilling with auger bit

Twist bit

A twist bit is a general-purpose bit ideal for drilling smaller holes.

Figure 2.42 Twist bit

Types of screwdriver bit

It is important to select the correct screwdriver for the job in hand. 'Star' or 'cross' head screws are not all the same and using the wrong type for the screw you are using can cause the screwdriver to slip. This, in turn, can make the screw head cam out or even cause an injury!

The three main types of screw head are:

- slotted

- phillips

- pozidriv.

Slotted   Phillips   Pozidriv

Figure 2.43 Three types of screw head

ACTIVITY

Find out why you might use certain types of screw heads for certain jobs.

Tool maintenance

It is very important to maintain your tools. They should be rust free and, very importantly, must be sharp. Blunt tools are dangerous as they take a lot more force to use. Blunt tools will also give a poor finish. Tools can be sharpened using an oil or water stone.

There are usually two angles on a cutting edge: a **grinding angle** and a **honing angle**.

KEY TERMS

Grinding and honing angles: there are usually two angles on a cutting edge, the grinding angle of 30 degrees, and a honing angle of 25 degrees.

Sharpening stones

Combination stones are used with oil and need to be used carefully as they can become hollow. If a stone is not flat, it can ruin your tools. Combination stones have a fine cutting side and a coarse cutting side.

Figure 2.44 Selection of sharpening stones: a diamond stone and an oil combination stone

TOOLBOX TALK

Do not wear any clothing or jewellery that dangles when using a grinder. Long hair should be tied back.

Diamond stones are made from tiny diamonds bonded to a metal plate. These stones are generally used with water and kept flat.

A slow grinder (wet grinder) is a machine that slowly turns a wet stone. The stone sits in a bath of water that washes away the metal from the sharpened tool and keeps the tool cool. There are guides and clamps that hold the tool being sharpened.

TRADE TIP

When using a slow grinder, remember to take the stone out of the water when finished. If you do not, it will absorb the water and go out of shape.

Figure 2.45 Note the angle of the grind

Use of stones

Firstly, put a little oil on the tip of the blade and move it forward until a little oil squirts out from the front of the chisel. This means the cutting edge is against the stone.

Sharpen the angled part of the blade on the stone using a 'Figure of 8' motion. If using an oil stone, be careful not to use too much oil.

A Oil on chisel tip

B 'Figure of 8' action

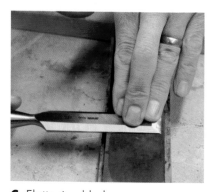

C Flattening blade

D Incorrect flattening of blade

Flatten the back of the blade; it is important to keep the blade flat.

If the blade is not kept flat, the blade will be damaged.

TRADE TIP

Backing off will ruin a chisel or plane iron. This happens when the back of the blade is not kept absolutely flat on the stone. This makes the blade impossible to sharpen, and the only remedy is to grind the affected area away.

TRADE TIP

The best way of telling how much blade is showing when setting a plane is to look down the bed from the front towards the handle into the light. Rubbing your fingers over the blade doesn't help and is dangerous.

 HAVE A GO...

Break down a smoothing plane. Remove the capping iron and backing iron as shown on page 39 and sharpen the blade. Put the plane back together in reverse order. To set the plane for depth of cut, use the adjustment wheel. You only want a little blade to show, just enough to be seen. It is likely that the blade will not be level with the bed of the tool, so adjust it using the adjustment lever.

Use of grinders

You will need to be properly trained before you can use a grinder.

Put the tool into the guide before turning on the grinder. If it is a grinder like the one shown in Figure 2.46 with a vertical stone, make sure that the stone is sitting in some water. Be sure that there is nothing touching the stone and that any dangly bits of clothing or long hair are tied back before turning the grinder on. Move the tool slowly from side to side, taking care to cover the whole face of the stone.

Figure 2.46 Slow grinder

Techniques

So far in this chapter, you have learned about the materials and tools needed for woodworking. In this second part of the chapter you will learn about some basic woodworking joints. This involves cutting, then assembling a joint and, sometimes, driving in a fixing to hold it all together.

Order of operations

It is best to mark out and cut joints in an organised sequence. This makes it easier and prevents you wasting time.

1	Prepare timber
2	Select face and edge
3	Mark out all joints
4	Chop mortices
5	Run tenons (do not cut shoulders)
6	Run rebates and moulds
7	Cut shoulders

Figure 2.47 Table of work sequence

Types of joint

First, take a look at some types of joint:

- butt joint
- halving joint
- mortice and tenon joint
- tusk tenon joint
- flooring joint
- studwork joint
- rafter joint
- housing joint
- dovetail joint
- scribe joint
- mitre joint.

Butt joint

A butt joint is the most basic of wood joints, but often it isn't a strong joint. One piece of wood is simply fixed to another piece with nails or screws. This joint is often used in studwork.

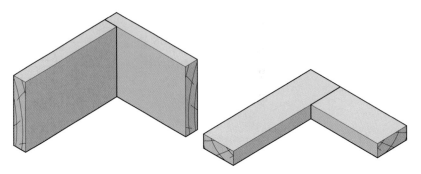

Figure 2.48 Butt joint

Halving joint

A halving joint is a fairly simple joint to produce. It is formed when joining two pieces of wood along its length or at corners.

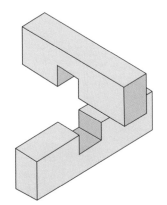

Figure 2.49 Halving joint

Mortice and tenon joint

A mortice and tenon joint is a very strong joint used for frames, doors and windows. Notice how the joint in Figure 2.50 is fixed using wedges. The wedges need to be well fitting, with the grain going in the correct direction or they could snap.

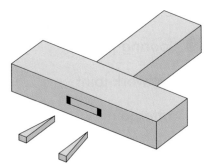

Figure 2.50 Mortice and tenon joint

Tusk tenon joint

Figure 2.51 shows a tusk tenon joint. Note the proportions of the joint; these proportions work whatever the dimensions of the **joist**.

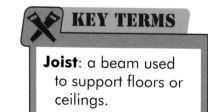

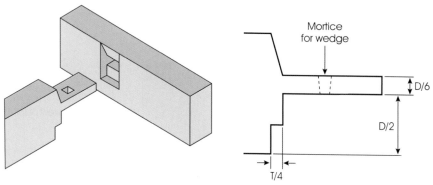

Figure 2.51 Tusk tenon joint

Flooring joint

Figure 2.52 shows a selection of traditional methods of jointing joists.

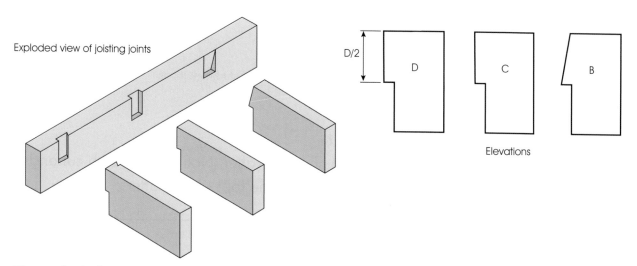

Figure 2.52 Flooring joints

Studwork joint

Figure 2.53 shows a selection of studwork joints, used in producing partition (internal) walls.

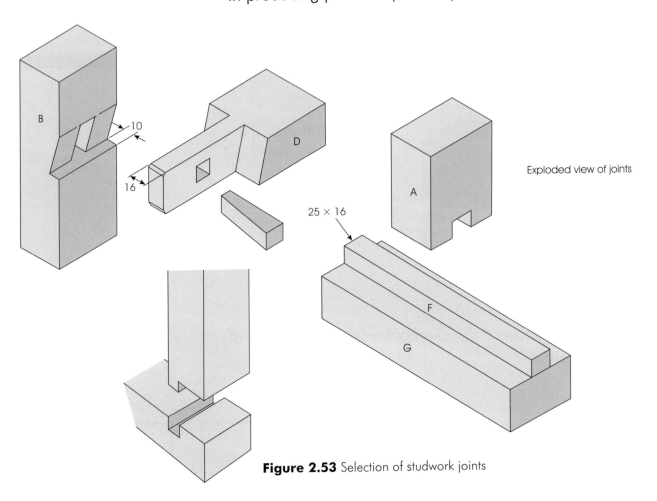

Exploded view of joints

Figure 2.53 Selection of studwork joints

Rafter joint

Figure 2.54 shows a rafter. Notice the plumb cut and birdsmouth joint at the foot of the rafter.

Figure 2.54 Rafter with plumb cut and birdsmouth

Housing joint

Figure 2.55 shows a housing joint. This is often used in door linings or for shelving.

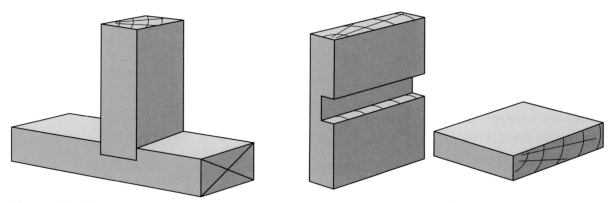

Figure 2.55 Housing joint

Dovetail joint

A dovetail joint is commonly used in furniture manufacture for items like boxes and drawers. It is decorative and strong, but time consuming to produce.

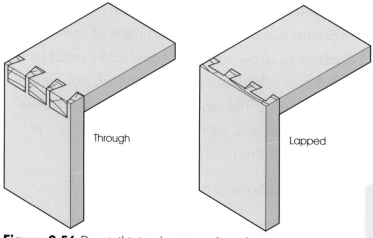

Through

Lapped

Figure 2.56 Dovetail joint showing tails and pins

Scribe joint

A scribe is used for internal joints on skirting.

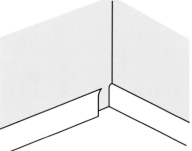

Figure 2.57 Scribe joint

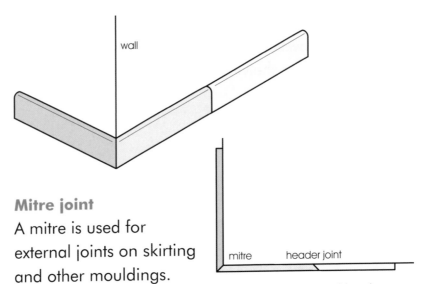

wall

Figure 2.58 Mitre and header joint

mitre   header joint

Mitre joint

A mitre is used for external joints on skirting and other mouldings. A similar joint can be used to join lengths of skirting. This is known as a header joint.

Marking out

Marking out joinery and cabinet making

Before marking out and making an item of joinery, it is best practice to draw a rod. Ideally, tasks need to be carried out in a logical sequence. Mark out all the joints before cutting. The only exception to this rule is when making dovetails, as it is customary to cut the tails first and mark the pins from the tails.

The series of operations is as follows:

- When you select your timber, make sure any defects such as knots are not on joints; this could make it difficult to cut the joint or make the joint weak. **Shakes** can be placed away from the **face and edge**.

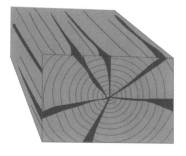
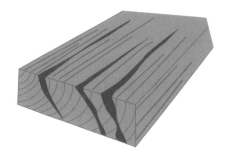

Figure 2.59 Shakes in wood

■ The face and edges will need to be marked first before marking anything else. When using the square and gauge, always mark from the face. When producing a frame, the face edges will all point inwards and the faces to the front. This ensures that you produce **stiles and rails** of the correct hand.

Figure 2.60 Partially assembled frame showing face and edge marks

☐ Now mark the joint positions including waste. The waste will be cut off later. Stiles or rails that are the same length can be marked at the same time; this saves time and ensures that they are exactly the same.

Figure 2.61 Marking out joints

☐ Use a gauge to mark the depth or position of the joints from the face.

☐ When marking out a dovetail, use a **dovetail ratio** of 1:6.

Figure 2.62 Gauging the joint

Figure 2.63 Dovetail being marked out with a pitch of 1:6

KEY TERMS

Dovetail ratio: this can vary. It is usual to use a ratio of 1:6 for softwood and 1:8 for hardwood. Some people use 1:7 for all dovetails. 1:6 means 1 unit across the shoulder to 6 units along the joint. So, for a 60 mm deep joint, you would come in 10 mm along the shoulder.

TRADE TIP

When using a mortice gauge, set it to the chisel you are about to use. This is because chisels often vary in size. A chisel marked 10 mm might be over or under by as much as 1 mm. The joint will not fit properly if the chisel doesn't exactly match your marking out.

■ Hatch waste with a pencil. This ensures you cut on the right side of the line.

Figure 2.64 Hatching waste

Marking out carpentry

As with joinery or cabinet making, sometimes it is a good idea to draw out your markings first. This can be done to **scale** if your markings are too large to fit on to paper or a plywood sheet.

When marking out studwork or joists, the measurements are usually given as **centres**.

KEY TERMS

Scale: many things are too large to draw on a piece of paper, so to get them to fit the dimensions are reduced. For instance, an item 2 m long will not fit on paper but a 200 mm item will. So the item has been reduced by a tenth, or 1:10. This is a scale of 1 to 10.

Centres: joists and rafters are spaced from centre to centre.

QUICK QUIZ

What is a shake?

a) A milky looking piece of wood

b) Where a branch has grown out of a piece of wood

c) A split in a piece of wood

d) A cut in a piece of wood

TRADE TIP

When marking out centres, try to keep the rule or tape in the same place. This makes it much more accurate. For example, if you mark out 400 mm centres without moving the tape, you will mark 400 mm, 800 mm, 1200 mm, and so on. If you move the tape you could gain or lose 1 or 2 mm each time. Over 10 centres, you could be 20 mm out!

EXPANDING YOUR SKILLS

When marking out rafter plumb cuts and birdsmouth joints, you will need to determine the length of the rafters and the pitch of the plumb and seat cuts. There are several methods of doing this.

The roof can be drawn out on a sheet of plywood or MDF. For this, you will need to mark the measurements given and join up the points. This will give you the angles of the plumb and birdsmouth cuts. These angles can be transferred to the timber using a sliding bevel. The length will be marked on a line that is two-thirds of the depth of the rafter shown.

A Drawing section through roof

B Taking bevels using sliding bevel

C Marking out on to rafter

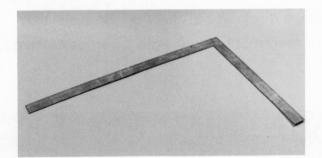

D Roofing square

Another method is to use a steel square. There are several types of square available, and you will have to follow the manufacturer's instructions and tables provided for the square you use.

Cutting

Preparing timber

Prepare square timber with a plane. This should be done with a jack plane as a smoothing plane is short and will not give a very straight cut. Make sure that the plane is sharpened and set up as explained on page 29.

A First, plane up the face.

REMEMBER

Always keep your work area clear. Put away tools you no longer need, throw away shavings and offcuts and keep the floor clean – not just when you finish, but as you work. This will help you to work more efficiently and safely.

B Then, plane up the edge. Check that it is square with the face.

C Gauge the correct thickness and depth using a marking gauge and then plane down to a line.

Cutting to length

This is usually done with a tenon saw.

As soon as the cut has started, move your thumb away. Notice how the saw is being held, with one finger on the side of the handle. This improves the control of the saw. Follow the line you have squared all around the timber. Take care when finishing the cut as it might split.

Figure 2.65 Notice the position of the thumb when starting the saw

Tee halving joint

This technique can also be used for making a housing joint:

A Cut down the sides of the housing. This is usually done with a tenon saw.

B Using a sharp chisel, remove the middle of the joint. Notice the 'roof' shape.

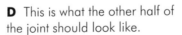

C Now remove the centre, taking care not to go right through the material as this will cause it to split. Use your gauge line to put your chisel in to finish the housing.

D This is what the other half of the joint should look like.

The other half of the joint is cut using a similar technique to cutting a tenon (see Figures A–C on page 54).

It doesn't matter if the joint is a little hollow but if there is a high point inside, the joint will not fit. A long narrow housing can be finished with a hand router.

Morticing

There are several methods of morticing. Below is one method. Use a bench cramp to hold down the material, with a scrap piece of wood to prevent damage to the timber.

REMEMBER

Do not cut towards your hands or body. Keep your fingers out of the way. If you slip, the chisel is likely to go into you!

A Start in the centre, leaving about 5 mm each end of the mortice.

B Work your way down, making sure that the waste is pulled out of the joint each time.

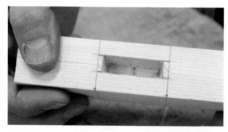

C When you get halfway through, turn the wood around. Don't go all the way through, you will split the wood.

D When the two mortices meet, clean out any remaining waste with a blunt instrument (such as a rule).

E Clean out the ends of the mortice.

  HAVE A GO...

Mark out a piece of wood as shown above using a marking gauge. Remember to use the face to mark from. Practise cutting the mortice. Remember not to go all the way through! Chop it out halfway and turn the wood around.

Cutting tenons

On smaller tenons or halving joints, this will be done with a tenon saw. If it's a larger tenon, you can use a hand saw. It is important that you move your hand out of the way when the cut is started. If the saw slips out while you cut, it could injure your hand quite badly.

If your joinery product has a rebate, you should run the **cheeks** first, run the mould and then shoulder the tenon after rebating.

A Cutting a cheek. Lean the timber away from you and cut from corner to corner.

B Turn the timber around and cut from the other corner.

C Finish the cut.

Notice the way the cut follows the hatched side of the line – this is the waste. The saw has a **kerf** and if you cut down the middle of the line, the tenon or halving joint will be the wrong size. Leaving too much waste after the saw cut will mean that you have to clean up the joint with a chisel. This is time consuming and difficult to do. Practise to see if you can cut to the line – it will save you time in the long term.

D Shoulder the joint.

Figure 2.66 Kerf

KEY TERMS

Kerf: the thickness of the cut removed by a saw.

HAVE A GO...

Mark out a piece of wood as shown using a marking gauge. Remember to use the face to mark from. Practise sawing down the side of a line so you can just see where the line was, but there is no excess waste left on the tenon.

EXPANDING YOUR SKILLS

Sometimes shoulders are not square to the edge of the timber. This is the case when a rebate does not continue through the joint. This is known as a splayed shoulder.

TRADE TIP

If you leave too much waste on the shoulder, you will have to **pare** it. This is difficult and time consuming. See if you can practise to cut on the line.

KEY TERMS

Pare: to chisel across the grain. Commonly done to clean up shoulders. A special type of bevel-edged chisel, known as a paring chisel, is designed for this operation.

Figure 2.67 Cutting a mitre. The mitre can be marked and cut freehand or be cut in a guide known as a mitre box.

Mitres and scribes

Mitres and scribes are used at the corners of moulded timber, such as skirting and architraves. Mitres are used for external joints (around the outside of a corner) and scribes are used for internal joints (around the inside of a corner).

The shape of a scribe can be cut as follows:

A Mitre the moulded part as shown.

B Cut the flat part of the scribe first, with a tenon saw, from the bottom. Then cut the curved part with a coping saw. The scribe is slightly undercut, meaning the face of the joint fits tightly.

C Assemble the scribe.

Chamfers, rebates and grooves

The parts of a joinery item are often shaped to improve its appearance or to allow panels or glass to be fitted to it. This is known as moulding.

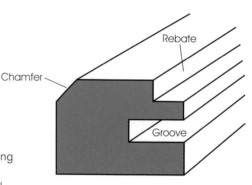

Figure 2.68 Chamfer, rebate and groove

A A chamfer is made using a jack plane. Mark the depth of the chamfer, and plane down to the line.

B A rebate is made using a rebate plane. Set the fence and stop at the required size. When setting, remember to measure to the blade not the bed of the tool.

C A groove can be made using a plough plane.

TRADE TIP

When marking a chamfer, do not use a gauge as it will leave a line you cannot plane out.

Cutting curves

Curves are cut with a coping saw. As the curve will need cleaning up, cut about 1 mm past the line on the waste side. When cutting, look at where the blade is going rather than the frame of the saw, as sometimes the blade will twist. Be sure to keep the blade under tension or it will snap.

The curve will need to be cleaned up using a spokeshave. Notice the points for the thumb on the handle. Remember to use the right type of spokeshave and make sure it is really sharp.

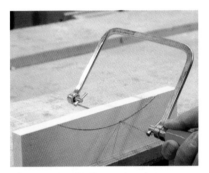

Figure 2.69 Cutting with a coping saw

Figure 2.70 Finishing with a spokeshave

Dovetails

When making dovetails, it is best to cut the dovetails first and then mark the pins (or housings) from the dovetail.

A Cut the tails.

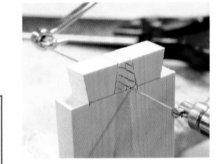

B Remove the waste with a coping saw. You will have to clean this up with a sharp chisel.

C Mark the pins using your tails.

Assembly

Before you assemble a job, make sure you have everything you need. Have all the components ready before you start: cramps, wedges, fixings, glue and tools. Make sure that the cramps are laid out flat and not twisted.

Cramping

Cramping is very important when assembling a job. It makes sure that the joints come up tight, and holds the joints together while fixing them. The outside wedges are put in first – this pushes the rails tight into the mortices and ensures there are no gaps around the inside of the frame.

Figure 2.71 Notice how the cramps are positioned as close as possible to the joints without covering them. Note the blocks of wood that prevent bruising from the cramps.

Screws

Screws are commonly used for fixing. However, it is important to pre-drill them before driving them in as they may not fix properly and the wood might split.

Nails

Nails can split the wood, especially near the edge if you are not careful. Pre-drilling or blunting nails first can prevent this. Use a hammer to drive in the nail. A **nail punch** can be used to drive the nail below the surface.

KEY TERMS

Nail punch: a tool used to punch a nail below the surface.

TOOLBOX TALK

Once the nail has started to go in, keep your hand out of the way in case the nail punch slips off and hits your fingers!

TRADE TIP

The face of the hammer should be clean and free of dirt such as glue. If the hammer is dirty, it is unlikely to drive the nail correctly and may slip off causing the nail to bend.

Wedges

Wedges must be cut accurately – and remember to get the grain going the correct way!

Figure 2.72 Cutting a wedge

Figure 2.73 Using a wedge, showing clearance

Figure 2.74 Checking for square using a squaring rod

When assembled, check the frame for square and twist. A square frame will measure the same across each corner, so this is a good way to check. Look down the side of the frame from one side to the other; the two pieces should be in line. If they are not, the frame is twisted (in wind). A frame that is out of square should be squared up by applying pressure to the long diagonal before the glue dries. A slight twist can be put right by gently twisting the frame the other way until it springs back flat. Be careful not to twist it too hard or you might break it!

TRADE TIP

When squaring up or taking out twist, often you will hear a little 'click' when enough pressure has been applied. This means that the joints have moved slightly.

Finishing

When the frame is assembled, a joiner will need to finish it with a smoothing plane. Make sure the smoothing plane is set fine and is very sharp. Using a plane properly takes a lot of practice. Upon completion, the job can be lightly sanded, but this is only a final finish as it must be planed first.

Carpentry jobs do not need planing, but make sure any screws or nails are finished below the surface.

IN THIS CHAPTER you have learned about:

- ☐ the different types of wood trades
- ☐ woodworking materials
- ☐ woodworking tools
- ☐ tool maintenance
- ☐ techniques.

Remember that producing work to a good standard takes time and you will get quicker as you get better. Take time to get the job right – speed will come later!

CHECK YOUR KNOWLEDGE

REVISION QUIZ

1. What would be the best tool to use to cut a curve?

2. Name two different chisels and say what each is used for.

3. What is a kerf?

4. How long would a 2 metre measurement be if drawn at a scale of 1:10?

5. What type of saw has a back to the blade?

6. What tool would you use to mark angles?

7. How many teeth per 25 mm does a dovetail saw have?

8. What is a rod used for?

9. What is a rebate plane best used for?

10. Name the main two types of timber.

3 Trowel occupations

Bricks, stone and mortar were used to build the ancient world and many of these buildings still exist. Even materials we use today, like concrete, were used by the Romans.

This chapter looks at three trowel trades:

- ■ Brickwork and blockwork
- ■ Tiling
- ■ Plastering.

In this chapter you will learn about:

- ■ trowel occupations
- ■ trowel occupation materials
- ■ trowel occupation tools
- ■ tool maintenance
- ■ techniques.

The information in this chapter covers skills and knowledge to complete units: 005, 006, 009, 106, 110, 111, 116, 117, 118, 119, 205, 207 and 208.

Trowel occupations

Bricklayers/blocklayers

A bricklayer and blocklayer erect walls using bricks and blocks, and generally mortar for the joints. These walls can be single or cavity, and can feature quoins, piers and lintels. (These terms are explained on pages 71, 79 and 65.)

Plasterers

A plasterer applies render and plaster to walls to give a good finish. Rendered walls will have a rough finish, whilst plastered walls should be smooth.

Tilers

A tiler will hang and lay tiles, usually in places where there is a lot of wear or in places where there is water. A tiler will cut, fix and **grout** the tiles.

Trowel occupation materials

Bricks

Bricks are commonly made from one of three different materials: clay, concrete or calcium silicate. Bricks are smaller than blocks. The standard brick measures

REMEMBER

Bricks and blocks are heavy, and cement and lime are hazardous. Therefore, these must be used with care; if in doubt, ask first.

KEY TERMS

Grout: the filler between tiles that seals the joints and provides a decorative finish.

Figure 3.1 Types of brick

215 mm x 102.5 mm x 65 mm, but they do vary in size slightly because of the firing process when they are manufactured.

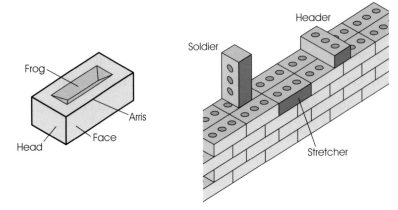

Figure 3.2 Brick showing frog, arris, header and stretcher face

Blocks

Blocks are made of **concrete**. There are many types of block, but the most commonly-used ones are dense concrete and lightweight **aircrete** blocks. Lightweight blocks are easier to use and can be cut with a saw. Aircrete blocks have excellent thermal properties. Standard blocks are 440 mm x 215 mm on the face and come in several thicknesses: 75 mm, 100 mm, 150 mm, 200 mm and 225 mm.

KEY TERMS

Concrete: a mixture of cement and aggregate (aggregate is small lumps of stone).

Aircrete: a mixture of cement, sand and power station waste. It has a lot of air in it and because of this it is lightweight.

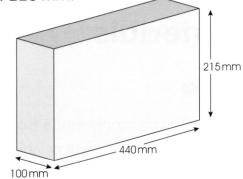

Figure 3.3 Dimensions of a concrete block

Other materials

■ Mortar is a mixture of sand and cement with water, although in many training workshops it is made from sand and lime so it can be reused. It is applied with a trowel and finished with a float.

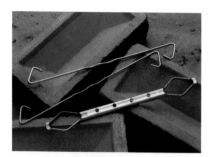

Figure 3.4 A selection of wall ties

■ Plaster is a powder that is mixed with water and then applied to the wall. There are different grades of plaster – some designed as a first coat, some as a top coat. This is trowelled up to give a smooth finish.

■ Damp proof course (DPC) is a waterproof membrane that prevents damp rising up through a wall.

■ Wall ties are used to hold two walls together when built side by side. There are many types of wall ties.

Figure 3.5 Wall ties in a cavity wall

■ Lintels are used to bridge gaps in walls, for example over a door or window opening.

Figure 3.6 PCC (pre-cast concrete) and steel lintel

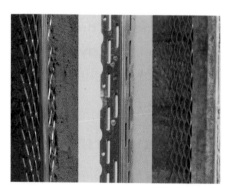

Figure 3.7 Corner beads (left and centre) and stop bead (right)

■ Beads are used when rendering or plastering. They are made from **galvanised** steel or stainless steel. This prevents the bead rusting in damp conditions.

KEY TERMS

Galvanised: this is when steel has been given a thin coating of a metal called zinc. This prevents rusting.

Other materials

Plasterboard is a material commonly used for ceilings or walls. It can be fastened to studwork with nails or screws, or stuck to brick or block walls using adhesive. The joints can then be filled and sanded or the board can be plastered. There are quite a few sizes and thicknesses of plasterboard available. The smaller sizes are easier to handle but cover less area, so the resulting wall has more joints.

Figure 3.8 Insulation being held in place with a clip on the wall tie

Building regulations require that insulation is built into a wall to prevent heat escaping. Insulation is lightweight and needs to be kept in place using clips.

Cavity closures are made from plastic and are used to finish off the **cavity** at any openings in the wall.

Figure 3.9 Cavity closure in position

KEY TERMS

Cavity: the gap between two layers (or skins) in a wall that prevents damp crossing over from the outer wall to the inner wall. This is where the insulation is placed.

Tiling materials

Tiles

Tiles come in a variety of sizes and colours. They can be made from clay or sometimes stone. Wall tiles are about 6 mm thick; floor tiles are about 8 mm. Floor tiles are usually bigger than wall tiles. Tiles are finished with a shiny coating known as glaze.

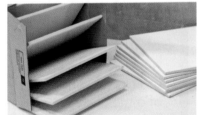

Figure 3.10 Tiles

Grout

Grout is used to fix and fill between the tiles.

Spacers

Spacers are used between the tiles when hanging or laying them. The spacers maintain an even gap between the tiles.

Figure 3.11 Spacers

Trowel occupation tools

Trowels

A trowel is a tool that comes in many shapes and sizes. It has many uses and a bricklayer or blocklayer will use a variety of trowels when working. (See also 'Tiling equipment' on pages 74–5.)

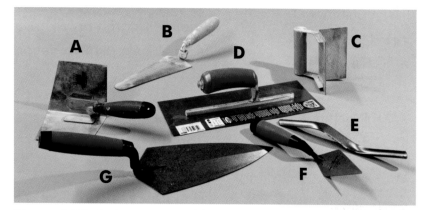

Figure 3.12 **A** plastering trowel for internal corners; **B** gauging trowel; **C** plastering trowel for external corners; **D** plastering trowel; **E** jointer or bucket handle; **F** pointing trowel; **G** bricklaying trowel

Brick trowel

A brick trowel (laying trowel) is a large diamond-shaped tool about 260 mm long, although there are much bigger trowels available. This is the most important tool to the bricklayer or blocklayer and will get continual use. Therefore, if buying a brick trowel, it is worth getting one that is comfortable to hold.

REMEMBER

A wall that has lots of gaps in the joints between the bricks is weaker than one where the joints are full.

/////////////

  **QUICK QUIZ**

What are the following tools used for?

a) Pointing trowel

b) Brick trowel

c) Jointer

Pointing trowel

A pointing trowel is a small trowel used to fill in joints between bricks or blocks when the joints are not full.

Gauging trowel

A gauging trowel has a rounded end and is ideal for measuring, small mixing jobs and patching in.

Plastering trowel

A plastering trowel is used to apply render and finish plaster. There are trowels available to finish into corners.

Jointer

A jointer (commonly known as a bucket handle) is used to give joints a decorative finish.

Cutting tools

Hammer

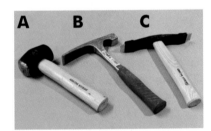

Figure 3.13 A Club hammer; **B** brick hammer; **C** comb hammer

■ A club hammer is used for hitting cold chisels or bolsters. It can have a handle made of wood, metal or fibreglass. There are different weights available: 1 kg, 1.5 kg and 2 kg.

■ A brick hammer is used for the rough cutting of bricks. The cutting edge should be sharp.

■ A comb hammer is similar to a brick hammer and has blades set into its cutting edges. When these wear down, they can be replaced. A comb hammer is used for cleaning up cuts and shaping bricks or blocks.

Bolster and cold chisel

■ A bolster chisel is used for cutting bricks and blocks. It has a wide blade that has been hardened. The striking end is softer to prevent the metal shattering when in use. However, this causes the end to mushroom, which can be hazardous. How to deal with this is considered on page 77.

Figure 3.14 Bolster and cold chisel

■ A cold chisel is smaller than a bolster and is hardened in a similar way. It is used for smaller cutting jobs, like removing small areas of plaster or cleaning up cut ends.

Saw

■ A block saw can be used to cut lightweight concrete blocks very neatly. It looks like a carpenter's saw, but has hard tips fixed to the teeth.

Figure 3.15 Saw for lightweight block

ACTIVITY: IDENTIFY

Identify the four tools below and explain what each one is used for.

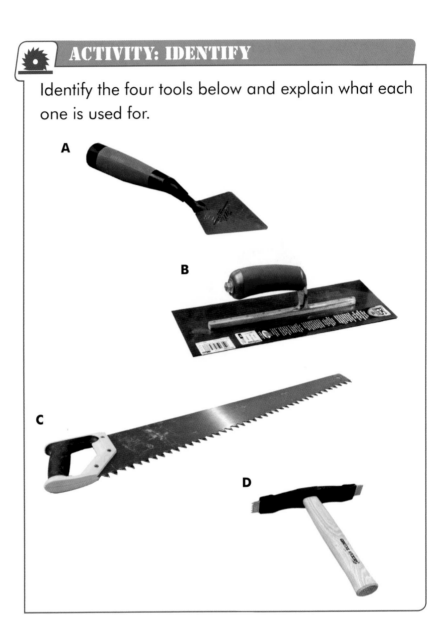

A

B

C

D

Measuring and setting out equipment

Walls need to be built neatly with straight lines so that they do not lean. Not only does this look better, it also makes the wall stronger. Let's take a look at some tools used to measure, set out and build straight walls.

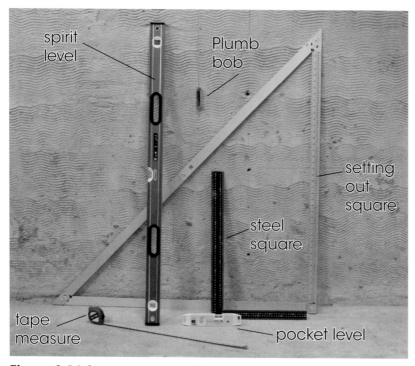

Figure 3.16 Some measuring and setting out equipment

A steel square is used for checking that bricks or blocks have been laid square at a **quoin**.

A tape measure is used for measuring. It's important to keep the tape measure dry and clean. If sand gets into it, the tape will rust and become worn very quickly.

A spirit level is used to check for **level** and **plumb**. Try to keep the vials clean (the clear parts where you can see the bubbles).

A pocket level (boat level) is used for smaller levelling jobs.

A plumb bob is used to plumb lines.

KEY TERMS

Quoin: a corner in a wall.

Level: a perfectly horizontal line. This is measured using a spirit level.

Plumb: a perfectly vertical line. For example, the weight of a plumb bob is pulled down by gravity and pulls the line tight; this line is now plumb. To determine a vertical line is to 'plumb' a line.

Figure 3.17 Spirit level in use

A line with pins or corner blocks is used to lay blocks or bricks. The line is stretched between two points (often the corners) and **ranges** between these. It is important to keep this line tight. If it droops, so will your wall.

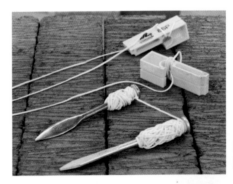

Figure 3.18 Corner blocks, line and pins

Figure 3.19 Line in use

A gauge rod is used to maintain the **gauge** of the wall and has lines marked on it representing the height of each course, including the mortar bed.

KEY TERMS

Gauge: this refers to the thickness of each course or layer of bricks or blocks in the wall.

Figure 3.20 Gauge rod in use

A pinch rod is used for checking diagonals; this can also be done with a measuring tape. If a square or oblong has sides that are the same as the side opposite and the diagonals measure the same, then the corners must be square.

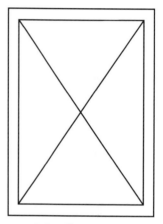

Figure 3.21 Illustration of square with diagonals

Figure 3.22 Ranging pegs in use

Datum pegs (ranging pegs) are used to transfer levels over rough ground. They are driven into the ground using a lump hammer.

REMEMBER

When driving pegs, take care not to hit your hands. Make sure your fingers are nowhere near the top of the peg.

Tiling equipment

Adhesive trowel

An adhesive trowel is used to apply adhesive to the tile or wall. The trowel can be combed, toothed or grooved and often will have a combination of these patterns. The purpose of the different patterns is to apply differing thicknesses of adhesive as required. The flat edge of the trowel is used to apply the grout on completion of fixing the tiles.

Figure 3.23 Adhesive spreaders

Tile cutter

A tile cutter uses a hardened wheel to score the surface of the tile. A lever is then used to break the tile along the score line.

Figure 3.24 Tile cutter

Tile saw

A tile saw used for tiling is similar to a coping saw used for woodwork, except that the blade is specially designed to cut tiles.

Figure 3.25 Tile saw

Scribe

A scribe is a hardened steel pin or cutter used for free-hand cutting.

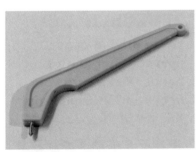

Figure 3.26 Scribe

Grout float and sponge

A grout float and sponge is used to finish off the grouting.

Figure 3.27 Grout float and sponge

Other equipment

Spot board

A spot board is made from a square of plywood and used to put mortar on before use. The spot board will need to be level and positioned near to where you are working.

Float

A float is used to finish render. It can be made of wood or lightweight foam plastic.

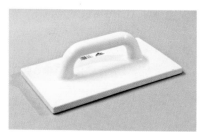

Figure 3.28 Float

Hawk

A hawk is a hand-held spot board and is useful for working plaster or mortar before applying it to the wall.

Figure 3.29 Hawk

Mixing auger

A mixing auger is used to mix plaster. It can be hand powered or put into a drill.

Jamb brush

A jamb brush is used for dusting off.

Figure 3.30 Mixing auger

Figure 3.31 Jamb brush

Straight edge

There are several types of straight edge available; these can be made of wood or aluminium. A straight edge can be used to check range and is very useful for rendering and plastering.

Figure 3.32 Types of straight edge: aluminium feather edge and darby.

Tool maintenance

Over time, trowel occupation tools become worn and damaged. The most common defects that need maintenance are hammer handles, comb hammer tip replacement and mushroomed striking faces on bolsters.

A hammer handle can be replaced like this:

A Remove old handle.

B Fit new handle.

The tip on a comb hammer gets worn over time and can be replaced like this:

C Replace tip on comb hammer.

REMEMBER

Look after your eyes. Make sure any cutting tool you use doesn't have a mushroomed head. A piece might break off when you hit it!

Mushroomed heads can be removed by the percussion method. Be sure to wear goggles when doing this:

D Place mushroomed head on a hard surface.

E Strike mushroom from underneath.

Techniques

Bonding

Bricks and blocks need to be overlapped to make sure the wall is strong; this is known as bonding. Look at Figure 3.33 on page 78, which shows two walls – the first one is not bonded, the second is bonded. The

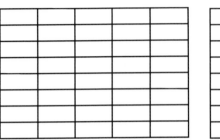 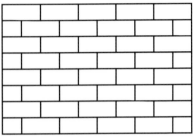

Non-bonded wall   Bonded wall

Figure 3.33 Non-bonded and bonded walls

Half-brick and one-brick thick walling: a stretcher bonded wall is the same thickness as a brick cut in half widthways, so it is called half brick thick. The English bonded wall is as thick as a full brick is long, so it is called one brick thick. Walls can be thicker than this, for example one and a half brick thick.

TRADE TIP

Dry bond your wall first. This will enable you to work out your bond and cuts prior to laying the bricks and blocks.

second wall is much stronger because the bricks are locked together in a pattern. See how the load on the wall is spread over a wide area. Long vertical joints in walls are avoided for this reason.

Figures 3.35 and 3.36 show two examples of common bonds. Notice the English bonded wall is thicker than the stretcher bonded wall. The stretcher bonded wall is **half-brick thick**, whilst the English bonded wall is **one-brick thick**.

Figure 3.34 Stretcher bond   **Figure 3.35** English bond

It can be difficult to work these bonds out in your head, so it is good to lay the bricks out dry (without cement) on the ground first. This is known as dry bonding.

Look at the quoins in Figures 3.37 and 3.38 – two in brick and one in block. Notice the cut bricks and blocks that are needed to maintain the bond.

Figure 3.36 Two brick quoins in stretcher and English bond

Figure 3.37 A block quoin

A pier is a 'tower' of bricks or blocks. This can be attached to a wall or free standing.

Figure 3.38 Pier

Cutting

Bricks and blocks need to be cut in order to maintain the required bond. Bricks and dense concrete blocks can be cut using a lump hammer and bolster chisel. Lightweight blocks can be cut using this method too, but there are also hard point saws, for example block saws, available that will do the job just as well (see page 69).

KEY TERMS

Closer and half bat: a brick cut in half lengthways is known as a closer; a brick cut in half across its width is known as a half bat.

Figure 3.39 Cutting a **closer** and a **half bat**

Laying bricks and blocks

You have learned how to bond and cut bricks and blocks. Now we will look at how to lay them. Each brick and block will be laid on a bed of mortar, and it takes some skill to get them level, straight, neat and even. When building the wall, it is best to build up the ends first and then fill in the middle. Figures A–H show bricks, but laying blocks is very similar.

A Be sure to set up your spot board close to the work area, and place your mortar on the board. With a brick trowel, cut off some of the pile and form a roll of mortar.

B You will want to put down your first layer (course) of bricks. You should place the end bricks first. These will need to be level and straight across the face with each other.

C The mortar bed for the first course on the floor is laid out. The mortar bed is spread out with the tip of the trowel.

D Next, pick up a brick. Apply mortar to the end of the brick. This is known as 'perping' the brick. The vertical mortar joints in a wall are known as perpends.

E The bed for a brick laid on top of other bricks (i.e. the second course upwards) is laid like this. Again, notice how it is spread with the tip of a trowel.

F Lay the brick and tap with the handle of the trowel. It is important that the brick has the correct thickness of mortar bed or gauge. The brick must be level, plumb and laid to the line.

G Remove excess mortar with the trowel.

H Range can be checked with a straight edge placed along the front of the wall. It's also important to check the brick is laid square and the wall is level.

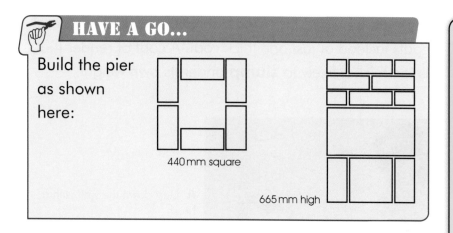

Pointing

It is important that bed and perpend joints are full and well finished. This ensures that the bricks and blocks are all well supported and bonded to each other, making the wall stronger. It also stops water getting into the wall and causing damp problems. If the joints are not full, they will need pointing. A pointing trowel and jointer are used for this.

Two basic joint finishes are flush and half round (bucket-handle).

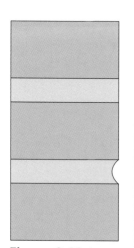

Figure 3.40 Joint finishes: flush and half round

Rendering and plastering

Applying render

Render is made from sand and cement and applied to a wall using a trowel. It is important that the area to be rendered is clear of dust and damped down. If it is dry and dusty, the render will not stick and is likely to fall off.

Slump: when the weight of the plaster or render starts to pull the coat down the wall and it bellies out at the bottom..

It should be noted that render must be applied in several coats instead of just one thick coat. A coat of render that is too thick is likely to **slump** under its own weight.

A Dust down the wall with a brush

B Wet the wall

C Apply the first coat using a hawk and trowel. Be sure to have your spot board nearby.

D Use a feather edge board to remove excess render.

E The first coat needs to be 'scratched'. This provides a good key for the next coat.

F Float finish the second coat.

G If a finishing plaster is to be applied, then the finished render is given a scratched finish using a special float.

EXPANDING YOUR SKILLS

Bell casts are decorative and protect whatever is below them from the weather. They are formed using special beads.

A Cut and fit beads. The bead should be rested on to a batten fixed to the wall.

B Apply first coat. Don't forget to scratch it to provide a good key.

C Finish bell cast.

Applying plaster

Plaster is a much finer material than render and provides a smooth finish. There are different types of plaster. One-coat is used for general work where a lot of thickness is required; board finish is used for finishing plasterboard.

Plaster is applied using a skimming trowel. To provide a very smooth finish, plaster must be trowelled again once the plaster has begun to **go off**.

KEY TERMS

Go off: to set, go hard.

A After preparing the wall, mix the plaster in a bucket with an auger.

B Using a gauging trowel, check the consistency of the plaster. It should stick to the trowel as shown and be lump free.

C Apply plaster to wall in two coats.

D As the plaster begins to set, trowel off the wall to a very smooth finish. A brush is used to apply water during this process.

EXPANDING YOUR SKILLS

Finishing a wall using the 'dot and dab' method:

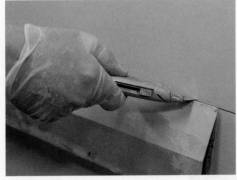

A Measure and then cut plasterboard along face with knife, right through the paper layer.

B Snap board.

C Cut paper from behind.

D Dab surface with adhesive.

E Press board into place, checking for plumb.

Tiling

Tiles are used around sinks, baths and basins to provide a water-resistant surface. They are stuck to the wall or floor using grout.

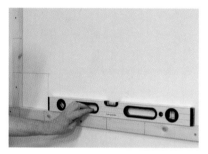

A Marking out wall to be tiled. Use a level and plumb bob to make sure the tiles are fixed level and plumb. Fix battens to fix the tiles to.

B Marking out floor to be tiled. Notice how the cut tiles are placed at the edges of the room and how the cut tiles are equal at each wall.

C Scoring tile. Remember to score just once.

D Make cuts for things like switches and pipes with a tile-cutting coping saw.

E Snap tile.

F Apply adhesive and fix tile. Remember to use spacers.

G Grout on completion. Push grout into joints.

H Remove excess grout and then finish with sponge.

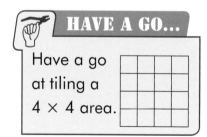

HAVE A GO...

Have a go at tiling a 4 × 4 area.

IN THIS CHAPTER you have learned about:

- ☐ trowel trades
- ☐ trowel materials
- ☐ trowel tools
- ☐ techniques.

CHECK YOUR KNOWLEDGE

REVISION QUIZ

1. What is a jointer?

2. What is a half bat?

3. Name two brick bonds.

4. What does 'half-brick thick' refer to?

5. What is a bell cast?

6. Name two joint finishes.

7. What would you use a float for?

8. How many times do you score a tile when cutting?

9. What is a mushroomed head?

10. What does 'range' refer to?

Plumbing

Plumbing with lead and copper pipes goes back thousands of years. It was used by the Egyptians and Romans. Today, plumbing materials are made from copper, plastic and steel.

In order to complete your Basic Construction Skills course, you will need to learn some pipework skills. These include basic plumbing skills, such as bending and joining plastic and copper tube, and fitting sanitary ware and rainwater systems.

In this chapter you will learn about:

- the plumbing trade
- basic plumbing materials
- basic plumbing tools
- techniques.

The information in this chapter covers skills and knowledge to complete units: 004, 007, 108, 109 and 204.

The plumbing trade

A plumber's job is to fit pipes for hot and cold water supplies. They will also fix internal items such as sinks, baths, showers and basins, connecting these items to waste services. As well as these jobs, a plumber will also fit gutter and rainwater systems.

Basic plumbing materials

Modern plumbing uses a variety of materials made from metals and plastics.

Figure 4.1 Plastic pipe

Fittings

Plastic tube is now very commonly used for hot and cold water supplies, as well as central heating. It is easy to cut, bend and install. However, plastic doesn't look as nice as copper, so tends to only be used for work that is hidden. Plastic tube is commonly available in 15 and 22 mm.

Figure 4.2 Copper tube

Copper tube is a lot stiffer than plastic and will need to be bent using tools. It looks nicer than plastic, and so it is used for work that is seen. Copper tube is available in 15, 22 and 28 mm.

Plastic waste pipe is used to carry waste water away from things like sinks, basins and baths. Waste pipe is available in 32 and 40 mm.

Figure 4.3 Waste pipe

Push-fit fittings are used to join pipes together and there is a wide variety available, as can be seen in Figure 4.4. These fittings are simple and easy to use; the pipe just pushes into the fitting and the joint is made (be sure to push the pipe right in). Push-fit fittings can be reused.

TRADE TIP

Make sure that the cut pipe end is clean, as push-fit fittings rely on a thin rubber washer inside to stay watertight and it is easy to damage.

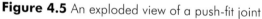

Figure 4.4 A selection of push-fit fittings: (from left to right) straight connector; reducing coupler; equal tee; elbow; reducing tee

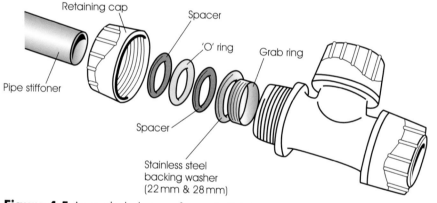

Retaining cap — Spacer — 'O' ring — Grab ring — Pipe stiffener — Spacer — Stainless steel backing washer (22mm & 28mm)

Figure 4.5 An exploded view of a push-fit joint

Figure 4.6 Olive

Compression fittings are also used to join pipes. The watertight join is made by inserting the pipe with an **olive** and tightening the nuts with a spanner. These joints should be used with a jointing compound containing a fibre called hemp or PTFE (polytetrafluoroethylene) tape on the threads to make sure the joint is watertight. Compression fittings can be reused if the olive is replaced.

KEY TERMS

Olive: a copper or brass washer used with a compression fitting. It is important it is not damaged.

Figure 4.7 Close-up view of a compression fitting

Capillary fittings are used to join pipes using solder and a blow torch. Yorkshire fittings have solder already in them, making them more convenient than standard capillary fittings.

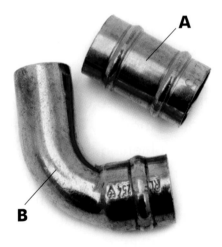

Figure 4.8 Capillary fittings: **A** end feed; **B** Yorkshire or solder ring. The fittings shown above for push-fit are also available for **capillary joints**.

Taps should be fitted to a sink or basin before the sink is fixed into position. A common method of connecting taps is by using flexible tap connectors.

Figure 4.9 Flexible tap connector

Solder is made from metal that has a low melting point. It is melted to join the tube together. This process is called soldering.

Flux is used when soldering to make sure there is a good joint.

Figure 4.10 Solder wire

Figure 4.11 Flux

TOOLBOX TALK

Solder should be lead free, as lead is harmful to health.

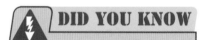

DID YOU KNOW

Flux prevents the metal combining with oxygen (causing corrosion) and helps the solder flow and stick.

Figure 4.12 Soil pipe

A soil pipe is used to take waste away from the house. It is 110 mm in diameter.

A pan connector is used to connect a **WC** to the soil pipe. It has a rubber gasket and fins that adapt to the pan.

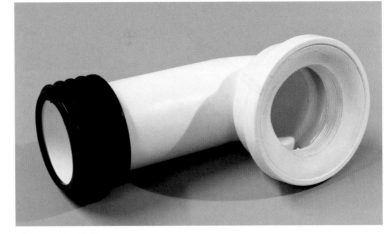

Figure 4.13 Pan connector

Waste pipe push-fit fittings are very simple to use but can be quite a tight fit. Therefore, a lubricant is recommended to help push the joint together.

Figure 4.14 A toilet

Figure 4.15 Waste pipe

Fig 4.16 Lubricant

ACTIVITY: IDENTIFY

Name these three fittings:

A

B

C

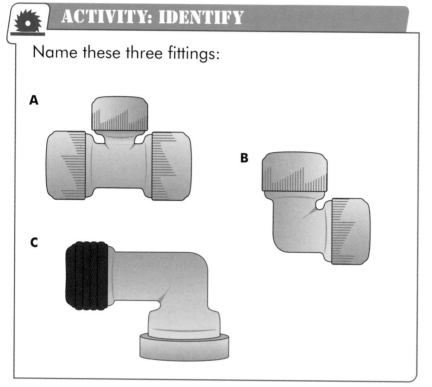

DID YOU KNOW

Although Thomas Crapper is credited with inventing the flushing toilet, evidence of flushing toilets goes back to Neolithic times!

Flush and filling mechanisms in toilets can be separate or combined. These control the flush and how the cistern fills. An overflow protects against the cistern overfilling if the filling mechanism goes wrong.

Figure 4.17 Flush and filling mechanisms

Figure 4.18 A hand basin

A trap is fitted to sinks, baths and basins. This blocks smells coming up from the sewerage system by trapping some water at the bottom.

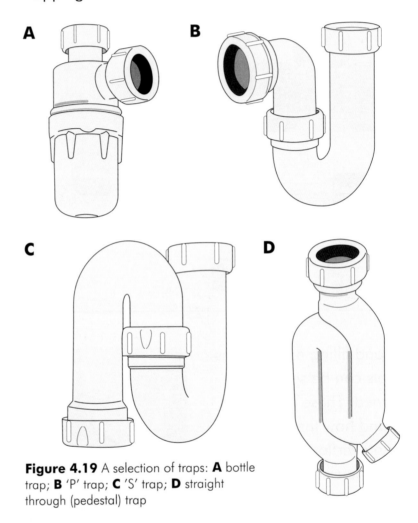

Figure 4.19 A selection of traps: **A** bottle trap; **B** 'P' trap; **C** 'S' trap; **D** straight through (pedestal) trap

Rainwater goods

Rainwater goods can be made from a wide variety of metals, including cast iron, aluminium and copper. In this book, we will consider the most common – plastic. There is a huge variety of colours, shapes and sizes coming from many manufacturers.

A gutter catches the rainwater as it comes off the roof and carries it to an outlet. The fall should be approximately 10–17 mm for every 6 m length of gutter (600:1 to 350:1).

Figure 4.20 A gutter

A union is used to join lengths of gutter.

The downpipe takes the rainwater down the wall.

A swan neck is needed to get from the fascia to the wall. It is made up of two bends and sometimes a short straight piece.

Figure 4.21 A union

Figure 4.22 A downpipe

Figure 4.23 A swan neck and shoe

TRADE TIP

A gutter should not have too much fall because then the water will run away quickly without carrying away any debris. Over time, grass and plants will grow in this debris.

A shoe fits on the bottom of the downpipe where the water comes out into the drain. Brackets are used to hold the gutter. Clips hold the downpipe to the wall.

Figure 4.24 Brackets

REMEMBER

The spacing between brackets should not be more than 1m.

Figure 4.25 Running outlet and stop end

A running outlet connects the gutter to the downpipe and the stop end stops the water running off the wrong end of the gutter.

Basic plumbing tools

Plumbers have a lot of tools, many of which are quite heavy. Here is a selection:

A hacksaw is used for general cutting.

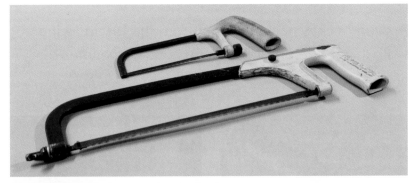

Figure 4.26 Hacksaw and junior hacksaw

Figure 4.27 Pipe cutters (for copper and for plastic)

There are many types of pipe cutters available. Some are designed for plastic and some for copper.

A **burr** removing tool is used to remove any burr from the cut end of the tube.

An adjustable spanner is used to tighten compression fittings.

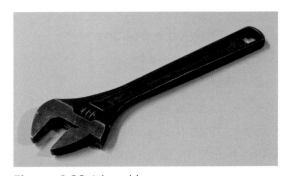

Figure 4.28 Adjustable spanner

KEY TERMS

Burr: a sharp piece of metal left remaining after cutting a pipe. This needs to be removed as it could damage the fitting.

A pipe bender is used to form neat corners without having to use fittings.

Figure 4.29 A pipe bender

Wrenches are used for holding pipes and tightening joints.

Figure 4.30 Wrench

A file is used to remove sharp edges from cut pipes.

A blow torch is used to solder capillary joints.

A fireproof solder pad or tile should be used when soldering joints to protect the surrounding area from scorching.

Figure 4.32 Blow torch

TOOLBOX TALK

Pipe-bending equipment is heavy. Take care when handling.

Figure 4.31 A file

TOOLBOX TALK

The end of a blow torch will remain hot for quite a while after you have finished using it.

ACTIVITY: IDENTIFY

Identify the four tools below and explain what each one is used for.

A

B

C

D

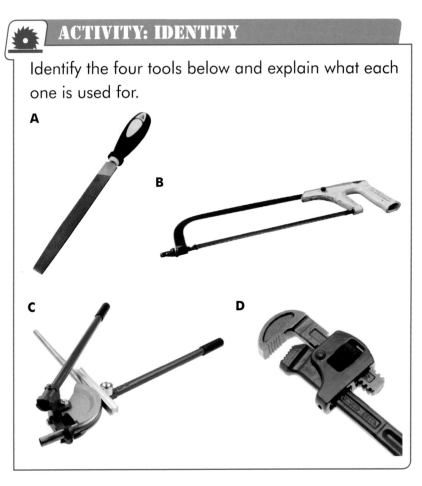

You will also need the following tools (which are all explained in Chapters 2 and 3):

- ☐ Screwdriver.

- ☐ Battery drill.

- ☐ Spirit level.

- ☐ Line.

- ☐ Plumb bob.

- ☐ Tape measure.

- ☐ Hand saw.

Tool maintenance

Tools should be kept sharp and in good condition. Pipe cutters become blunt over time. When this happens, the cutting wheel should be replaced. Hacksaw blades should be replaced when necessary. Equipment should be lubricated correctly.

Access equipment

A plumber would need access equipment from time to time (for example, to put up rainwater goods). This is covered in more detail in Chapter 5 on page 121 and is also considered in Chapter 1 on page 10.

Plumbing techniques

Plumbing involves installing items like sinks and toilets (sanitary ware), and then connecting them up to waste and water supply services. This involves cutting, bending and joining pipes. You will also need to fix these pipes to the wall neatly to give a good appearance.

Installing sanitary ware

Things like sinks and toilets need to be fitted to the wall and floor before you can plumb them in. It is easier to fit taps and mechanisms before fixing sanitary ware and sinks.

A Installing flush and filling mechanism

B Installing WC pan

C Assembly of WC pan and cistern

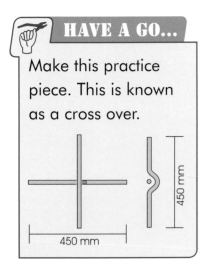

HAVE A GO...

Make this practice piece. This is known as a cross over.

450 mm

450 mm

D Fitting taps before fixing wash hand basin

Cutting

Copper and plastic tube can be cut using a hacksaw or a pipe cutter. There are special cutters for plastic. Waste pipe is cut using a hacksaw.

Clean the pipe if needed and remove any burr left after cutting the pipe.

REMEMBER

It is very important that the tube is clean before jointing.

A Cutting copper pipe with pipe cutter

B Cutting waste pipe with hacksaw

C Removing burr

Bending

Accurate bending of pipes is done using a pipe bender. You will need to be accurate with your measurements.

A Measure the bends.

B Set up the bender. Make sure it is adjusted correctly, as the pipe can be crushed (known as 'throating') or given a rippled effect instead of a smooth bend.

C Make the bend. Remember to overbend slightly, as the pipe will spring back a little.

Figure 4.33 Connecting push-fit fitting. Remember to push the tube right up to the stop and make sure that you have not damaged the o-ring inside.

Jointing using push-fit

Push-fit fittings are very simple to use and make excellent watertight joints. However, the joints tend to be quite bulky compared to other types of fitting, so they tend to get used for plumbing that is out of sight.

Jointing using compression

Compression joints are a little more difficult to use than push-fit. You may need to use hemp with jointing compound or PTFE tape to ensure a good watertight joint.

A Slide nut down pipe.

B Fit olive and check it is not damaged.

C Apply hemp or PTFE tape to threads if required.

D Assemble joint.

E Tighten joint.

HAVE A GO...

Make this practice joint, which involves push-fit and compression joints.

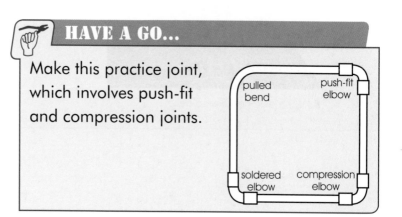

pulled bend

push-fit elbow

soldered elbow

compression elbow

EXPANDING YOUR SKILLS

Jointing using capillary solder fittings gives a very neat and watertight joint. This method is often used for work that is on show.

A Prepare tube making sure it is clean. You will not get a good joint with a dirty tube.

B Apply flux.

C Push fitting on to tube.

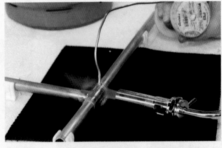

D Solder joint with blow torch. Remember to use a fireproof mat as shown. You will get a watertight joint when you see the solder run all around the joint. Don't overheat the joint – the copper will blacken and might give a poor joint.

E Wipe away excess solder

IN THIS CHAPTER you have learned about:

- ■ basic plumbing
- ■ some plumbing materials
- ■ basic plumbing tools
- ■ techniques.

CHECK YOUR KNOWLEDGE

REVISION QUIZ

1. Name two types of capillary fittings.

2. Name two different rainwater items and say what each is used for.

3. What is a blow torch used for?

4. What is a WC?

5. Why must you remove the burr from the end of a cut pipe?

6. What is an olive used for?

7. What is a swan neck?

8. Name three methods of cutting pipe.

9. What can you put on to the thread of a compression joint to ensure a good joint?

10. What two defects can happen if a pipe bender is under- or over-tightened?

5 Painting and decorating

After the other trades have left, the painters and decorators come in and provide the finished product. Any defects will be repaired. The standard of the finishing is what the customer sees.

This chapter will look at painting, wallpapering and applying hardwood finishes.

In this chapter you will learn about:

☐ the decorating trade

☐ materials used in decorating occupations

☐ tools used in decorating occupations

☐ decorating techniques and skills.

The information in this chapter covers skills and knowledge to complete units: 008, 112, 113, 114, 115 and 206.

The decorating trade

A painter and decorator will prepare surfaces and paint walls, ceilings, doors, windows and mouldings. Other jobs include **hanging** wallpaper and applying waxes, varnishes and polishes to wood.

In some cases, a good finish will depend on how well the decorator has prepared the item, or area, to be painted. Decorating is not just about applying paint or hanging wallpaper; it is also about removing old coatings and papers if needed, and then carefully filling and sanding the surface to ensure that it is clean and smooth ready for the finish.

KEY TERMS

Hanging: refers to pasting paper on to a surface.

TRADE TIP

Good decorating is mostly about preparation. It often takes a lot more time to prepare the surface than to paint it.

Materials used in decorating occupations

Paints and finishes

There are different types of paint that a decorator will use and, depending on the condition of the area to be painted, several layers of different types of paint may be needed. It is not possible to just apply gloss paint on to bare wood and get a good finish. In addition, primer or undercoat used with no top coat will not leave an attractive finish either. Using different layers of paint is known as a paint system. We will look at some of the different types of paint available; some have to be used as part of a system, others do not.

Tool maintenance

Brushes and rollers need careful cleaning when they are finished with. Brushes left in water or white spirit will become messy and damaged over time. It is best to clean out the tools properly, so there is not a trace of paint left in the brushes when the job is complete. They also need to be stored properly.

Figure 5.21 Remember to clean out your brushes

Water-based paints are easily cleaned from brushes and rollers. Just rinse them out in water until all of the paint has gone and the water is running clear from the stock of the brush.

Oil-based finishes will need cleaning with white spirit.

TOOLBOX TALK

When using steps (a stepladder), remember to check for damage to the stiles (sides) and treads. Make sure the strings at the bottom are taut and that you are using the steps on level, stable ground that is free of rubbish.

TRADE TIP

Don't leave brushes in a jar. This will bend the bristles and ruin the brush.

TOOLBOX TALK

White spirit should be kept away from bare skin; it is not for washing your hands!

TRADE TIP

Clean your filler knife while the filler is still soft. It is very difficult to remove filler once it has set.

Decorating techniques and skills

Paint and wallpaper need careful application. It doesn't matter how good you are with a brush, you will not get a good finish if you don't prepare properly first.

Protect the surrounding area

Remember to always protect the area before starting a decorating job. There are different types of dust sheet you can use, such as cloth or polythene.

Surface preparation

It is very important that the surface should be properly prepared and cleaned before any work is started. Flaky old paint needs to be removed. Holes and cracks need to be filled, and imperfections then need to be sanded flat. A sanded surface will also provide a good **key** for a new coat of paint.

A Scrape the surface to be painted and remove any flaky old paint or plaster.

B Fill imperfections: use filler for holes and caulk for cracks in corners or along mouldings.

C Rub down the surface, flattening filler and providing a key for the new finish.

Applying paint to walls

Walls are usually painted with emulsion. It is usual to cover large areas with a roller and carry out the fine work, such as cutting in, with a brush. The wall may need several coats to cover a contrasting colour.

A Apply a watered-down coat of emulsion paint to new plaster. This provides a good key.

B Cut into corners and mouldings with a brush.

C Load roller using a roller bucket.

D Apply emulsion paint to the large areas with a roller.

Applying paint to wood

If painting bare wood, apply primer first. Remember to coat knots with knotting solution to prevent resin from the wood spoiling the finish.

Figure 5.22 Apply primer to bare wood

If applying undercoat, make sure that you do not put too much paint on the brush. Too much paint on the surface can produce runs and drips in the paintwork. Once dry, rub down the surface lightly and apply a second coat of undercoat.

Figure 5.23 Applying undercoat

Apply the top coat. Take care not to overload the brush, or apply too much paint, as this will cause runs. Too much paint on the corners could result in something called 'fatty edges', where the paint becomes very thick.

Figure 5.24 Applying topcoat

EXPANDING YOUR SKILLS

When painting in two colours, you will need to cover one area with masking tape. Take care to keep a straight line or the finished result will look untidy when you remove the tape.

A Cover one area with masking tape.

B Remove the tape carefully.

TRADE TIP

It is important to match the colour of the undercoat to the colour of the top coat. A light top coat will need a white undercoat; black paint will need a grey undercoat.

HAVE A GO...

Prepare and paint a small piece of board. Remember to fill in any cracks and sand the area.

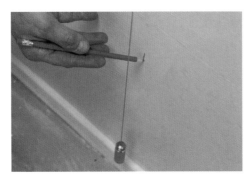

A Mark a plumb line down the wall with a plumb bob; the first piece of patterned paper will be hung to this line.

B Measure the paper and add about 100 mm to allow for trimming. When papering around a corner, allow about 10 mm overlap. It is very unlikely that the corner will be exactly plumb.

C When pasted, fold the paper like this. The paper may need to soak for a little while before applying it to the wall. With some modern papers, you have to paste the wall rather than the paper.

D Carefully apply the paper down the plumb line, use the wallpapering brush to smooth out the paper.

E Make a 'star' cut where there is a socket or switch.

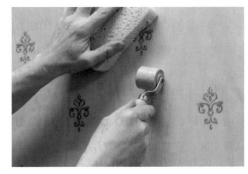

F Trim along the ceiling and skirting. Then, roll the joints using a sponge to remove paste.

Specialist wood finishes

Hardwood finishes are applied with brushes, rags and pads or buttons. These finishes are used for furniture and joinery, such as windows and doors.

A Rub down the hardwood surface first, taking care to sand along the grain.

B Apply oil with a rag or brush. Work with the grain. Each coat will build the finish.

C Varnish and stain can be applied with a brush. Be careful not to put too much on at once. Work with the grain.

D Apply French polish to the prepared surface using a rubber in a 'figure of 8' pattern. Keep building up the layers until the required finish is achieved.

French polish takes a bit of work to get right. It slowly builds up into a glossy finish.

TOOLBOX TALK

Oil and French polishes are flammable. A bin filled with rags that have been used for oil or French polish is a fire hazard. Dispose of rags carefully!

TRADE TIP

The rubber should be opened and polish must be poured into the rubber. The rubber should **not** be dipped into the polish.

IN THIS CHAPTER you have learned about:

■ the painting and decorating trade and types of jobs

■ decorating materials

■ decorating tools

■ a range of techniques and skills used.

CHECK YOUR KNOWLEDGE

REVISION QUIZ

1. What would be the best tool to use to cover a large wall area with emulsion?

2. Name two different brushes and say what each is used for.

3. What is a caulking gun?

4. Why should you not leave brushes in a jar?

5. Emulsion paint is available in what finishes?

6. How is French polish applied?

7. What is masking tape used for?

8. What do you need to think about when cutting patterned wallpaper?

The construction industry

6

Now that you have had an introduction to the skills you need for wood occupations, trowel occupations, plumbing and painting and decorating, what jobs can you train for?

In this chapter you will learn about:

☐ the construction industry

☐ the jobs available within the construction industry.

Craft and building operative

Jobs available under this heading include all the trades, general building operatives and specialist building operatives.

Craft operative job title	Role
Carpenter	Involved mainly on building sites with erecting timber frame walls, wooden floors, partitions, roofs and internal works, such as kitchens, stairs and skirting
Joiner	Works mainly on a bench and produces items of joinery, such as doors, windows and stairs
Bricklayer	Builds walls with bricks and mortar
Craft mason	In some areas, bricks are not in general use. A craft mason will use concrete blocks and mortar to build walls
Electrician	Installs wires and electrical systems into buildings
Painter and decorator	Decorates buildings. A decorator is often the last person on site, and leaves a finished product for the client
Plasterer	Plasters walls and ceilings, leaving a nice flat finish
Plumber	Installs pipes for water supplies, drainage and heating systems
Shopfitter	Fits out shops with shop fronts, units and shelving
Roofer	Installs roof covering in tiles or slates
Stonemason	Cuts and installs stone
Wood machinist	Uses machines to produce timber sections
Formworker	Installs formwork (otherwise known as shuttering) to pour concrete into

Building operative job title	Role
General building operative	Moves and mixes materials, uses plant (site vehicles such as dumpers and diggers) and generally helps out on site
Specialist building operative	Specialises in things like tiling or scaffolding

Glossary

Aircrete: a mixture of cement, sand and power station waste. It has a lot of air in it and because of this it is lightweight.

Auger bit: a tool for drilling larger holes in wood.

Bench hook: this is used to hold timber on the top of the bench whilst cutting shoulders.

Birdsmouth joint: a joint used in roof construction, connecting the rafter to the top plate of a supporting wall.

Burr: a sharp piece of metal left remaining after cutting a pipe. This needs to be removed as it could damage the fitting.

Cam out: this is where the screw head becomes damaged and it becomes very difficult to drive in or remove the screw.

Cavity: the gap between two layers (or skins) in a wall that prevents damp crossing over from the outer wall to the inner wall. This is where the insulation is placed.

Centres: joists and rafters are spaced from centre to centre.

Cheeks: the parts of timber at the sides of the tenon that get cut off.

Chuck: the part of the drill that the bit fits into. Most chucks do not need a key; older chucks need a key to tighten them up.

Closer and half bat: a brick cut in half lengthways is known as a closer; a brick cut in half across its width is known as a half bat.

Comb hammer: a hammer with blades set into its cutting edges. When these wear down they can be replaced.

Concrete: a mixture of cement and aggregate (aggregate is small lumps of stone).

COSHH: the Control of Substances Hazardous to Health.

Cutting in: a method for painting into corners using a brush.

Decant: to pour paint from one container to another, sometimes through a sieve to remove lumps.

Dovetail ratio: this can vary. It is usual to use a ratio of 1:6 for softwood and 1:8 for hardwood. Some people use 1:7 for all dovetails. 1:6 means 1 unit across the shoulder to 6 units along the joint. So, for a 60 mm deep joint, you would come in 10 mm along the shoulder.

Downpipe: pipe that takes the rainwater down the wall.

Eggshell: a satin or semi-gloss finish to painted surface.

Face and edge: the best two sides of the timber (and from where square and gauge lines are taken).

First fix: the carpentry jobs that are carried out before plastering.

Fitch brush: a type of paintbrush used for fine, detailed work.

Flux: substance used during soldering to prevent the metal combining with oxygen (causing corrosion) and to help the solder flow and stick.

Galvanised: this is when steel has been given a thin coating of a metal called zinc. This prevents rusting.

Gauge: this refers to the thickness of each course or layer of bricks or blocks in the wall.

Gauge rod: a type of measuring equipment used to maintain the gauge of a wall, with lines marked on it representing the height of each course, including the mortar bed.

Gloss: a highly shiny or reflective painted surface.

Go off: to set, go hard.

Grinding and honing angles: there are usually two angles on a cutting edge, the grinding angle of 30 degrees, and a honing angle of 25 degrees.

Grout: the filler between tiles that seals the joints and provides a decorative finish.

Half brick and one brick thick walling: a stretcher bonded wall is the same thickness as a brick cut in half widthways, so it is called half brick thick. The English bonded wall is as thick as a full brick is long, so it is called one brick thick. Walls can be thicker than this, for example one and a half brick thick.

Hanging: refers to pasting paper on to a surface.

Hardwood: timber that comes from a deciduous tree. Examples of hardwood trees are oak, ash and mahogany.

HASAWA: Health and Safety at Work Act – the law in the UK that deals with health and safety.

Hawk: a hand-held spot board used for working plaster or mortar before applying it to a wall.

Joist: a beam used to support floors or ceilings.

Kerf: the thickness of the cut removed by a saw.

Kettle: a plastic or metal container used to decant paint from a large tin to make it easier to transport and use.

Key: paints will not stick to a surface that is too shiny. A slight sanding will provide a suitable surface.

Knots: a knot is where a branch has grown inside the timber. Knots can be 'dead', which means they can drop out. They can be difficult to work with and can cause weakness in a joint. Often these areas contain a lot of resin that can come through paint and spoil the finish (known as bleeding).

Level: a perfectly horizontal line. This is measured using a spirit level.

Line and pins: a line with pins is used to lay blocks or bricks. The line is stretched between two points and ranges between these.

Matt: a dull finish that is not reflective.

Microporous: allows the wood to breathe and is excellent for outside jobs.

Mortice and tenon joints: a mortice is a hole cut into timber. This allows a tenon formed on a second piece of timber to fit into it. This strong joint can be wedged or dowelled.

Mortice chisel: a very thick, strong chisel used to make mortices in timber and to form mortice and tenon joints.

Nail punch: a tool used to punch a nail below the surface.

Olive: a copper or brass washer used with a compression fitting. It is important it is not damaged.

Pare: to chisel across the grain. Commonly done to clean up shoulders. A special type of bevel-edged chisel, known as a paring chisel, is designed for this operation.

Perpends: the vertical mortar joints in a wall.

Personal protective equipment (PPE): equipment worn to protect the user from things like dust, sparks and splinters.

Plough plane: a plane used to cut grooves

Plumb: a perfectly vertical line. For example, the weight of a plumb bob is pulled down by gravity and pulls the line tight; this line is now plumb. To determine a vertical line is to 'plumb' a line.

Plumb bob: a piece of equipment used to plumb lines.

Pointing trowel: a small trowel used to fill in joints between bricks or blocks when the joints are not full.

Push-fit fitting: a simple type of fitting used for joining pipes. The joints tend to be bulky so this is usually for plumbing that is out of sight.

Quoin: a corner in a wall.

Range: a straight line between two points. You will need to make sure your work is ranged, so that the face of your finished wall is flat.

Rebate: a square cut out along the length of the timber.

Repeat: some wallpaper patterns match straight across or halfway down the pattern.

Risk assessment: a form that is filled out highlighting all the risks involved with a certain job and how to deal with those risks.

Rod: a full-scale workshop drawing showing all the required detail.

Scale: many things are too large to draw on a piece of paper, so to get them to fit the dimensions are reduced. For instance, an item 2 m long will not fit on paper but a 200 mm item will. So the item has been reduced by a tenth, or 1:10. This is a scale of 1 to 10.

Second fix: the carpentry jobs that are carried out after plastering.

Shake: a split in the wood. There are several types of shake.

Shellac: a natural resin that is used in wood finishes.

Silk: a slightly shiny surface, i.e. with the appearance of silk.

Slump: when the weight of the plaster or render starts to pull the coat down the wall and it bellies out at the bottom.

Smoothing plane: a short plane used for cleaning up work.

Softwood: timber that comes from a coniferous tree. Examples of softwood trees are pine, redwood and spruce.

Solder: a substance used for / method of joining using metal which has a low melting point.

Stiles and rails: stiles are the upright parts of a door; the rails are the horizontal parts.

Subcontractors: workers not directly employed by the contractor, often known as 'subbies'.

Tenon saw: a type of saw with a back attached to the blade to make it stronger.

Tensioned: when a blade is kept tight. With a coping saw, this is done by turning the handle.

WC: water closet (toilet pan).

Index